Workshop on Materials and Engineering in Aeronautics (MEA2018)

IOP Conference Series: Materials Science and Engineering
Volume 476

Moscow, Russia
15 – 16 November 2018

ISBN: 978-1-5108-8217-1
ISSN: 1757-8981

Printed from e-media with permission by:

Curran Associates, Inc.
57 Morehouse Lane
Red Hook, NY 12571

Some format issues inherent in the e-media version may also appear in this print version.

Printed by Curran Associates, Inc. (2019)

For permission requests, please contact the Institute of Physics at the address below.

Institute of Physics
Dirac House, Temple Back
Bristol BS1 6BE UK

Phone: 44 1 17 929 7481
Fax: 44 1 17 920 0979

techtracking@iop.org

Additional copies of this publication are available from:

Curran Associates, Inc.
57 Morehouse Lane
Red Hook, NY 12571 USA
Phone: 845-758-0400
Fax: 845-758-2633
Email: curran@proceedings.com
Web: www.proceedings.com

Table of contents

Volume 476

Workshop on Materials and Engineering in Aeronautics

15–16 November 2018, Moscow, Russia

Accepted papers received: 19 December 2018
Published online: 18 February 2019

Preface

Papers

IOP Conf. Series: Materials Science and Engineering **476** (2019) 011001 doi:10.1088/1757-899X/476/1/011001

PREFACE

2018 Workshop on Materials and Engineering in Aeronautics (MEA2018) was held 15-16 November 2018 in Moscow Aviation Institute (National Research University). It was patronized by Moscow Aviation Institute (National Research University).

This Workshop was held 14 times from 2001 to 2017. Each was attended by over 20 participants who submitted their reports for publishing. The participants represented mostly the leading technical universities of Russia. The Scientific Committee hopes that the next year the international involvement would increase.

The (MEA2018) Workshop Proceedings include only 26 papers but the scope is actually huge. It covers materials & structure development, pilot interface, aerodynamics, aircraft parameter identification, system theory, control, aircraft technical state monitoring, robots, statistical classification and decision-making. This time we also include papers dealing with new tools for developing aerospace and other high-tech industries. In the year 2018 the Scientific Committee's confirms its traditional position not to limit the creativity of the participants. As a result all the papers are unlike one another and many of them contain original ideas and approaches. That is why our advice to a reader is not to judge the Proceedings by one or two papers but look through the most part of them.

On behalf of the Workshop Scientific Committee
the scientific editors of MEA2018 Proceedings:

Evgeny V. Nikulchev, Dr., Professor of MIREA – Russian Technological University, nikulchev@mail.ru

Oleg N. Korsun, Dr., Prof., Head of the laboratory, State Research Institute of Aviation Systems, marmottoduo@yandex.ru

IOP Conf. Series: Materials Science and Engineering **476** (2019) 011002 doi:10.1088/1757-899X/476/1/011002

Peer review statement

All papers published in this volume of *IOP Conference Series: Materials Science and Engineering* have been peer reviewed through processes administered by the proceedings Editors. Reviews were conducted by expert referees to the professional and scientific standards expected of a proceedings journal published by IOP Publishing.

IOP Conf. Series: Materials Science and Engineering **476** (2019) 012001 doi:10.1088/1757-899X/476/1/012001

Evaluation of the Efficiency of the Implementation of Public-private Partnership Mechanisms in the High-tech Industry

V V Artyakov[1], A A Chursin[2,3] and E S Andreeva[4]

[1] State Corporation «Rostec», Moscow 119048, Russia
[2] JSC Russian Space Systems ,Moscow, Aviamotornaya, 53, 111250, Russia
[3] Peoples' Friendship University of Russia, Moscow 117198, Russia
[4] Kazan National Research Technological University, Nizhnekamsk 423578, Russia

E-mail: esandreeva-nk@rambler.ru

Abstract. The article proposes a tool for assessing the efficiency of the implementation of public-private partnership mechanisms in the high-tech industry including aerospace. The paper presents a tool to assess the efficiency of public-private mechanism and the change of the efficiency estimation. Having the predicted pathway of the change in the efficiency evaluation, it is possible to monitor the real values of the indicator in the course of a project implementation, using the public-private partnership mechanism. Such monitoring will reveal significant deviations from a pathway, which can be of two types: deviations, excessing the actual values of an indicator over the predicted values and deviations, failing to achieve the real values of the predicted indicator

1. Introduction

The modern world economy is increasingly focused on a new type of economy - the knowledge economy. Today, the resources that determine the further development pathway of the national economy are neither raw materials and productive capacities, nor labor resources, but the unique competencies of corporations and employees. They play a crucial role both at the microeconomic and macroeconomic level.

The work [1] formulated the economic law on the interaction between the development of competences and consumer markets. Their interconnection leads to a spiral pathway and provides innovative economic growth, which suggests that the creation of unique competences increases resources in high-tech companies, leading to an abrupt occurrence of unique innovative technologies. Their use leads to the creation of fundamentally new products that form the requirements for new amenities and promote the emergence of new markets and economic growth that develops economic opportunities for creating new unique competences.

But to realize competencies in any highly competitive product or technology and bring the necessary result in the form of creating new consumer markets, it is first necessary to select the most promising projects for implementation, based on an assessment of their efficiency, and secondly, to ensure their investment attractiveness for commercial partners, since such projects are cost demanding.

Evaluation of the efficiency of projects and programs implemented in high-tech industries (including aerospace) is one of the most important components of their economic analysis. Indicators of the efficiency of projects and programs can be both absolute (for example, profit from commercial activities) and relative (economic, political and socio-social indicators). Evaluation of the efficiency of the implementation of programs and projects may indicate the significance of the results achieved, the extent of possible risk factors development and the degree of implementation of the activities, planned for a project or a program. The process of implementation of large projects is described in [2-4].

Public-private partnerships can play an important role in the implementation of major projects in high-tech industries [5,6]. Foreign experience shows the efficiency of PPP application in such industries. In this regard, it is necessary to have methods, based on the economic and mathematical modeling that enable to assess the prospects and efficiency of implementing a project, based on public-private partnership mechanisms. Simulation model is an effective tool for such analysis, which enables to take into account various economic factors, affecting the efficiency of PPP forms in the Russian Federation. Now many PP projects deal with the development of hi-tech competences creation what makes them especially risky [7-10].

Such models can be applied to solve a number of practical economic problems.

The proposed simulation economic and mathematical model provides opportunities to analyze economic risks when implementing large-scale projects in the form of PPPs including aerospace industry. The issues of practical risk assessment for a private partner in projects, based on the mechanism of public-private partnership are considered. The notion of threshold risk for a private partner is introduced and the way this concept can be used for practical risk assessment of the whole project is shown.

It is necessary to make a simulation model of a project implementation with the help of the public-private partnership, using the data on the amount of project stages financing by a state customer and a private partner, the predicted profit values, based on the results of a project implementation and random factors of the economy. Besides, it is necessary to consider various incentives (financial) that increase the investment attractiveness of a project for a private partner. Various economic indicators can be considered as the functions of the economic effect. For example, a ROI indicator can be such an indicator, equal to the ratio of the received income to the total investment volume).

2. Methods

To assess the efficiency of a project implementation, using the PPP mechanism at its various stages, let us offer an integral indicator (the ordinal number of a project implementation phase). Let us assume that the plan involves the implementation of a high-tech project or a program within a certain number of stages. Let us denote the stages in the form of a vector, which components can be represented as follows:

$$C = \begin{pmatrix} c_1 \\ c_2 \\ \\ c_n \end{pmatrix}$$

During the n stages of a project implementation, it is necessary to implement k activities. Let us designate them as follows:

$$A = \begin{pmatrix} a_1 \\ a_2 \\ \\ a_k \end{pmatrix}$$

The efficiency coefficient $M_i$is a measure that reflects the degree of a project implementation V_i and possible risks R_i: $M_i=M(V_i,R_i)$

On the basis of the quantities introduced, we can formulate the following optimal control problem:

$$M_i = M\left(V_{i,}R_i\right) \to \max,$$

that is, the problem of maximizing the efficiency coefficient.

Let us dwell on the procedure of these coefficients construction, taking into account the degree of a project implementation and possible risks that arise as a result of a project implementation.

As previously noted, a high-tech project involves a number of planned activities, their successful implementation impacts the success of a project as a whole. These measures, as a rule, can affect the efficiency coefficient in different ways. Thus, failure to meet the deadlines of various activities implementation can have different consequences. For example, failure to meet the deadline of any activity may lead to the failure of the consequent activities performance. At the same time, the delayed implementation of an activity that does not affect the implementation of other activities significantly does not lead to such large-scale negative consequences. Let us assign the weighting factor w_j to each activity A_j in order to consider such effect from failure to meet the deadline of activity implementation. The economic sense of the weighting factors involves the impact of the on-time implementation of a certain activity on the further progress of a project.

Let us denote the vector of weight coefficients as follows:

$$W = \begin{pmatrix} w_1 \\ w_2 \\ \\ \\ w_k \end{pmatrix}.$$

Weight coefficients must meet the following condition:

$$\sum_{j=1}^{k} w_j = 1.$$

The activities, carried out within the framework of a project can be implemented in stages (to recap, the calendar year is often considered as stages, but other division into stages is possible). In order to predict the changing efficiency evaluation from stage to stage, it is necessary to have the most probable estimates of the degree of implementation $q^i{}_l$ of a certain activity A_j upon completion of a stage C_i. The law of probability distribution of the degree of a certain activity A_j implementation within the project stage C_i may be written as a discrete law. Let the scale of this law suggest the following reference points of the implementation degree:

- 0.00 - no work was performed within the framework of an activity
- 0.10 - very low degree of an activity implementation
- 0.30 - low degree of an activity implementation
- 0.50 - average degree of an activity implementation
- 0.70 - high degree of an activity implementation
- 0.90 - very high degree of an activity implementation
- 1.00 - the activity is fully implemented

Let us combine these scale points into the following set: $l \in L = \{0, 0.1, 0.3, 0.5, 0.7, 0.9, 1\}$

The probability law of distribution of the degree of an activity A_j implementation at the moment of a stage C_i completion takes the form as in the table 1.

Table 1 The probability law of distribution of the degree of an activity A_j implementation at the moment of a stage C_i completion.

$L(A_j)$	0.00	0.10	0.30	0.50	0.70	0.90	1.00
$q^i{}_l$	$q^i{}_0$	$q^i{}_{0,1}$	$q^i{}_{0,3}$	$q^i{}_{0,5}$	$q^i{}_{0,7}$	$q^i{}_{0,9}$	$q^i{}_1$

The law of probability distribution of a variable $L(A_i)$ must meet the following formula:

$$\sum_{l \in L}^{k} q_l^i = 1.$$

Taking into account our arguments and using the obtained expressions, the formula for the evaluation of the degree of a project implementation V_i at the completion of a project stage C_i will be written as follows:

$$V_i = \frac{1}{k} \sum_{i=1}^{k} w_k \sum_{l \in L} l \cdot q_l^i .$$

Thus, the presented model has:

$$\max\left(\frac{1}{k} \sum_{i=1}^{k} w_k \sum_{l \in L} l \cdot q_l^i \right) = 1 .$$

It is the single value that this functional should take at the time of a whole project completion. In this case, we can say that activities within a project were implemented successfully.

The overall risk of a project (a project stage), implemented, using the mechanism of the public-private partnership, is assessed using the function $R=R(F, G)$ of two arguments that express the impact of risk factors on the winning functions of a state customer and a private partner. We can consider the standard value - the ratio of losses during the implementation of a project stage to the overall stage budget as a quantitative risk measure. By this means,

$$F = \frac{L_F}{S_F},$$

$$G = \frac{L_G}{S_G},$$

$$R = \max(F, G),$$

where L_F and S_F are the losses and the budget of a stage for a state customer, L_G and S_G are the losses and the budget of a stage for a private partner.

Taking into account the proposed formulas for calculating the indicator of implementation degree V_i of activities under a project (program) and the risk level R_i at a project stage C_i, the formula for calculating the efficiency factor Mi for this stage of a project implementation appears as follows:

$$M_i = M(V_i, R_i) = \left(\frac{1}{k} \sum_{i=1}^{k} w_k \sum_{l \in L} l \cdot q_l^i \right) \cdot (1 - R_i) .$$

In compliance with the planned deadlines for the implementation of activities and the insignificant demonstration of risk factors, the function of assessing the project implementation efficiency, using the public-private partnership mechanisms is ascendant.

The increase in the risk level has a negative effect on the growth of the efficiency evaluation, therefore

$$\frac{\partial M_i}{\partial R_i} < 0 .$$

Values for the efficiency evaluation of a project or a program implementation are in the interval $0 \le M_i \le 1$ and the limitary unit value should be achieved upon a project or program completion.

To monitor a project or a program implementation, it is possible to use the predictive pathways of changing estimates M_i.

To do this, all the parameters: risk assessment, coefficient estimation and efficiency index, necessary for calculation indicator M_i are evaluated at the stage of a project or a program development, using the method of expert evaluation. Estimates must be made for each stage of a project or a program. Next, you need to calculate the projected values of indicator M_i at each stage of a project implementation. As a result, you will get a time series $M_1, \ldots, M_N$ defining the predicted pathway of the indicator change. The projected value of the evaluation at the end of a project should correspond to the limit value.

3. Conclusion

Having the predicted pathway of the change in the efficiency evaluation, it is possible to monitor the real values of the indicator in the course of a project implementation, using the public-private partnership mechanism. Such monitoring will reveal significant deviations from a pathway, which can be of two types: deviations, excessing the actual values of an indicator over the predicted values and deviations, failing to achieve the real values of the predicted indicator. The first situation can refer to the excess of a project time limit over the planned one. The second situation may indicate problems and "stuttering" in the process of project implementation or the failure to implement the planned activities due to unforeseen circumstances. In this case, it is necessary to make management decisions that can return the values of the indicator to the predicted pathway or this pathway requires correction due to the changed conditions.

Acknowledgments
This paper was financially supported by the Ministry of Education and Science of the Russian Federation on the project No.26.1146.2017/4.6 «Development of mathematical methods to forecast efficiency of using space services in the national economy».

References

[1] Chursin A A, Shamin R V and Fedorova L A 2017 The mathematical model of the law on the correlation of unique competencies with the emergence of new consumer markets. *European Research Studies Journal* **3A** 39
[2] Cooke T 2013 Can knowledge sharing mitigate the effect of construction project complexity? *Construction Innovation* **13** 1.
[3] Dave B and Koskela L 2009 Collaborative knowledge management - a construction case study *Automation in Construction* **18** 7.
[4] Egbu C O 2004 Managing knowledge and intellectual capital for improved organizational innovations in the construction industry: an examination of critical success factors *Engineering Construction and Architectural Management* **11** 5
[5] Li Yaning Tang, Qiping Shen and Eddie W L 2010 Cheng A review of studies on Public-Private Partnership projects in the construction industry *International Journal of Project Management* **28** 7
[6] van den Hurk M, Brogaard L, Lember V, Petersen O H and Witz P 2016 National Varieties of Public–Private Partnerships (PPPs): A Comparative Analysis of PPP-Supporting Units in 19 European Countries *Journal of Comparative Policy Analysis: Research and Practice* **18(**1) 1-20 (DOI: 10.1080/13876988.2015.1006814)
[7] Chursin A and Tyulin A 2018 *Competence management and competitive product development: Concept and implications for practice* (Springer Intern Publ)
[8] Kashirin A, Semenov A, Ostrovskaya A and Kokuytseva T 2016 The Modern Approach to Competencies Management Based on IT Solutions *JIBC-AD-Journal of Internet Banking and Commerce* **01** 1-12
[9] Kashirin A, Semenov A, Ostrovskaya A, Kokuytseva T and Strenaluk V 2016 The Modern Approach to Competence Management and Unique Technological Competences *Quality – access to success* **17**(154) 105-109
[10] Shamin R, Chursin A, Kokuytseva T, Ostrovskaya A and Semenov A 2018 Modeling of Growth-Collapse Processes and Their Applications to Pricing Management *International Journal of Pure and Applied Mathematics* **118**(18) 3741-3746

IOP Conf. Series: Materials Science and Engineering **476** (2019) 012002 doi:10.1088/1757-899X/476/1/012002

To The Mathematical Modeling of Nonstationary Transonic Flows. Analytical Approach

A N Bogdanov

Institute of mechanics Lomonosov Moscow State University, 1, Michurinsky Pr., Moscow, 119192, Russia

E-mail: bogdanov@imec.msu.ru

Abstract. The transonic mode of gas flow has significant differences in comparison with both subsonic and supersonic regimes, combining both their general properties and specific features inherent in each of them, the possibility of penetration of weak disturbances upstream in subsonic flows. The behavior of small disturbances in subsonic and supersonic flows is qualitatively different. The paper proposes new analytical model for the mathematical modeling of unsteady transonic flows, including the effect of weak viscosity.

1. Introduction

The transonic mode of gas flow has significant differences in comparison with both subsonic and supersonic regimes, combining both their general properties and specific features inherent in each of them (for example, the existence of shock waves or Mach angles in a supersonic stream), the possibility of penetration (influence) of weak disturbances upstream in subsonic flows. The behavior of small disturbances in subsonic and supersonic flows is qualitatively different in a supersonic flow, these disturbances maintain amplitude along the characteristics towards which they propagate, in a subsonic flow, they attenuate at a distance from the source, capturing the entire subsonic flow zone. Such a variety of phenomena in their complex combination prooves [1] that in the study of transonic flows one can hardly rely on physical intuition or on considerations of plausibility.

The equations of stationary gas flow show (see, for example, [2]) that at the sound velocity of the flow the content model of the flow must be nonlinear. In this regard, the opinion was formed that the equations of flow in the transonic velocity range are essentially nonlinear and it is impossible to carry out linearization (the main method for simplifying the equations of the model).

2. Description of the mathematical model

A small parameter in problems of transonic currents is suggested by the physical peculiarity of the problem - the proximity of the speed of the flow to the sonic velocity. A simplified analysis of the equations of transonic flow is based on the transonic asymptotic expansions of the considered functions in series of the small deviations of the current value of the Mach number of the flow. In the case of a nonstationary transonic flow, such a simplification of the equations of motion was carried out by V. Lin, E. Reissner and S. Tsien, [3], which made it possible to reduce the original system to the only equation – Lin-Reissner-Tsien's equation (LRTs)

$$\left(K_\infty-(\gamma+1)\varphi_x\right)\varphi_{xx}+\varphi_{yy}-2\varphi_{xt}=0,\ K_\infty=\frac{1-M_\infty^2}{\varepsilon^{2/3}},\ M_\infty - \text{Mach number},\ \varepsilon\ll 1.$$

The LRTs-equation became basic in the study of unsteady transonic flows [4].

The development of the theory of gas flows historically developped from the study of subsonic flows to the study of supersonic flows, since such a course of events was determined by the capabilities and needs of aviation, but the generalization of the boundary layer model to pre-break modes for the first time was formulated for a supersonic flow by V.Ya. Neyland [5]. The resulting model was called, according to its structure, triple-deck. To emphasize the difference between the triple-deck model and the usual boundary layers, the name of a non-classical (self-induced pressure) boundary layer was also used for it.

The model was extended to the transonic mode by O.S. Ryzhov [6], on this model were obtained the main results in the study of non-classical transonic boundary layers. The initial version of the model showed the non-linearity of the flow equations in the upper deck. Further analysis found [7] that, in deriving the triple-deck model equations, it is possible to choose between maintaining the nonlinearity of the equations in the upper deck and taking into account the nonstationarity of the flow in the lower deck. The latter variant was used in determining the stability of the boundary layer to small perturbations, which is a linear problem. The LRTs equation was included in the model to describe the flow in the region of a non-viscous flow (upper deck).

The wealth of solutions obtained on a three-deck model for a transonic regime, however, overshadowed a very serious problem. Later it turned out [8] that the direction of research went the way leading away from reality - the classical model was singular, but this circumstance remained ignored. The LRTS equation, which has undoubted advantages (describes both supersonic and subsonic regions of the transonic flow, its non-one-dimensional, nonstationary and nonlinear character), however, has drawbacks that do not allow the propagation of nonstationary perturbations to be correctly described: this equation is degenerate hyperbolic equation and describes the propagation of non-stationary perturbations in the flow field only partly (only upstream). In this connection, a modified model was proposed for studying the problems of the theory of non-classical transonic boundary layer [8]. The modification of the model consists in the preservation (thus arising naturally) in the LRC equation, when it is deduced from the full equations for the potential, a singular (containing a small parameter) term of the transonic expansion with the second time derivative

$$\left(K_{\infty}-(\gamma+1)\varphi_{x}\right)\varphi_{xx}+\varphi_{yy}-2\varphi_{xt}-\varepsilon\varphi_{tt}=0$$

The equation obtained in this way is nondegenerate hyperbolic, its inclusion in the mathematical model of the flow gives a regular physical picture of the flow field and allows one to consider perturbations propagating in all directions.

3. Some results

Regularization of the model to describe the flow field when using the modified LRC equation is already visible from the wave front patterns from a point source of weak acoustic disturbances flown around by a transonic flow, defined by the equation [9]

$$K_{\infty}t^{2}+2t\left(x-x_{0}\right)-\varepsilon\left(x-x_{0}\right)^{2}=\left(y-y_{0}\right)^{2}\left(1+\varepsilon K_{\infty}\right)$$

This equation describes the propagation of perturbations in the flow field in all directions, the wave fronts from a point source of weak perturbations now have a physically justified closed form.

Further analysis of unsteady transonic flows using the modified LRC equation gives the following results. Dispersion relation obtained by studying the stability of a transonic flow, simulated by a three-deck model, for initial data

$$u_{0}=y,\; v_{0}=0,\; p_{0}=const$$

and disturbances of the form

$$f(y)\exp(\omega t+ikx)$$

has the form [5]

$$Ai'(\Omega)\frac{1}{\int_{\Omega}^{\infty} Ai(\zeta)d\zeta} = -\frac{(ik)^{5/3}}{\sqrt{\varepsilon\Omega^2 + (ik)^{1/3}\left(\Omega + (ik)^{1/3} K_\infty\right)}},\ \Omega = \omega/(ik)^{2/3},$$

where $Ai(\Omega)$ is the Airy function, the apostrophe is the differentiation of the function by its argument.

The bifurcation-like qualitative rebuilding of dispersion curves found in [10] when studying the stability of a nonclassical transonic boundary layer occurs when the transonic similarity parameter changes [11] without intersecting it - it does not exist at any value of the specified parameter, therefore the dispersion curves do not have branch points, which explains the difficulty analytic detection of such rearrangements.

Comparison of the results given by the classical and modified models showed that when using the modified three-deck model, the asymptotics for the vanishing viscosity of the lower branch of the neutral stability curve has a qualitatively different behavior from the classical model and is determined by the dependence

$$K_\infty = -\varepsilon\sigma^2 c_0^{\ 2} + \sigma c_0 - 1/\sigma^8,\ c = \omega/k$$

where σ is a constant characterizing the multiple difference of the neutral value of the wave number in the problem considered here from the one calculated for the subsonic boundary layer $k = k_0 = 1{,}0005$, $\omega = \omega_0 = 2{,}298$. It can be seen that the existence of a single neutral value for a given is an exceptional case (there are either two of them, or there is no such value at all [11]), indicating that the conditions of flow stability / instability are much more complicated than previously thought.

4. Conclusion

It also turned out that the previously considered unambiguous dependence of the perturbation frequency on the wavenumber (with the exception of a small range of small values) is not [11]. For applications, this means, in particular, that the problem of controlling the boundary layer with the help of a perturbation damper [12] becomes qualitatively more complex.

The proposed model can be effectively used in the study of other problems of non-stationary transonic aerodynamics.

Acknowledgments

The investigations were carried out according to research plan of the Institute of Mechanics of Lomonosov Moscow State University with partial financial support of Russian Foundation for Basic Research (projects 18-01-00793-a)

References

[1] Guderley K G 1962 *The theory of transonic flow* (Oxford etc.: Pergamon Press) p 344
[2] Cherny G G 1988 *Gas Dynamics* (Moscow: Nauka) p 424 (In Russian)
[3] Lin C C, Reissner E and Tsien H S 1948 *J. of Mathematics and Physics* **27**(3) 220-231
[4] Cole J D and Cook L P 1986 *Transonic aerodynamics* (Amsterdam, etc.: North-Holland) p 473
[5] Bogdanov A N and Diesperov V N 2005 *J. Appl. Math. Mech.* **69** (3) 394-403 (In Russian)
[6] Neyland V Ya 1968 *Abst. of Papers of the 3rd All-Union Conf. on Theor. and Applied Mech.* (Moscow: Nauka) 224 (In Russian)
[7] Ryzhov O S 1977 *Dokl. Academy of Sciences of the USSR.* **236**(5) 1091-1094 (In Russian)
[8] Ryzhov O S and Savenkov I V *J. Appl. Mech. Tech. Phys.* (2) 65-71 (In Russian)
[9] Bogdanov A N 1997 *J. Appl. Math. Mech.* **61**(5) 798-811 (In Russian)
[10] Ryzhov O S and Bogdanova E V *2006 IUTAM Symposium on One Hundred Years of Boundary Layer Research Solid Mechanics and Its Applications* **129**(2) 135-144
[11] Bogdanov A N, Diesperov V N and Zhuk V I 2018 *Comp Math. and Math. Phys.* **58**(2) 254-263
[12] Terent'ev E D 1986 *Doctoral Dissertation in Mathematics and Physics* (Moscow: Computing Center, Russian Academy of Sciences) 202 (In Russian)

IOP Conf. Series: Materials Science and Engineering **476** (2019) 012003 doi:10.1088/1757-899X/476/1/012003

Nonparametric Method for Aircraft State Prediction

A Chekin, Y Bondarenko, V Kosyanchuk and E Zybin

State Research Institute of Aviation Systems, Moscow, Russia

E-mail: chekin_a_u@mail.ru

Abstract. This paper is a part of a series of articles on nonparametric methods in dynamic systems theory. Here the authors propose a new method for direct dynamic problem solution based only on some past measurements of system input-output signals. The described nonparametric state prediction method needs no any priori information on system model parameters and does not require identification, training or statistical calculations. The methodological examples of nonparametric aircraft state prediction are presented.

1. Introduction

The aircraft state prediction is one of the core design problems of many functions of onboard and ground systems such as [1–4]: automatic flight envelope protection, aircraft takeoff and landing, terrain reference navigation, terrain awareness and warning, traffic alert and collision avoidance, 4D trajectory prediction, moving object guidance, and the others. The efficiency of the methods for aircraft state prediction depend mainly on available online information.

Traditional methods, based on aircraft mathematical models, work very well for own aircraft state prediction, when accurate model parameters are available, all input and output signals are measureable. But when enough information is not available the state prediction problem is dramatically more complex. The problems of nonstationarity and nonlinearity of aircraft complex dynamic models can partly be solved by inflight parameters identification methods [5–7]. But the problems of aircraft model unidentifiability in normal closed-loop operation conditions requires the additional controls exiting besides priori structural model information [8].

Moreover, traditional parametric methods can hardly be used for state prediction of other moving controlled objects because their mathematical models are very difficult to establish only by outer measurements. Using aircraft performance models with structures based on kinetic assumptions, also belonging to physics-based approaches, is also limited because most of these models based on ideal assumptions, rarely considered the real constraints, human behavior factors, and the intersection of trajectories [9, 10].

Among the comparatively new nonparametric prediction methods, that need no priori information on models parameters, the most widely used are statistical and artificial intelligence methods [11]. Compared with those parametric methods they are constructed with weak assumptions or even without assumptions. But these methods have high computational costs, need much sample size or learning time.

The authors propose a new method for nonparametric aircraft state prediction, that needs no any information on model parameters and does not require identification, training or statistical calculations. It significantly expands the applicability of nonparametric dynamic systems theory [12–15]. In this paper the simple own aircraft state prediction problem is solved, showing the principle of

IOP Conf. Series: Materials Science and Engineering **476** (2019) 012003 doi:10.1088/1757-899X/476/1/012003

the method. Further researches assume predicting of states of any observable moving objects with unexpected or expected command and control strategies given by pilots or flight management systems (known as aircraft intent).

2. Nonparametric aircraft state prediction problem formulation and solution

Let's an aircraft dynamics is described by a discrete state space model in the form the following parametric vector

$$x_{i+1} = Ax_i + Bu_i \tag{1}$$

or matrix

$$\begin{bmatrix} x_{i-h} & \dots x_i & x_{i+1} \end{bmatrix} = A\begin{bmatrix} x_{i-h-1} & \dots x_{i-1} & x_i \end{bmatrix} + B\begin{bmatrix} u_{i-h-1} & \dots u_{i-1} & u_i \end{bmatrix},$$

expressions, where *A*, *B* – eigen-dynamics and control efficiency matrices, *x* – state vector, *u* – control vector, *h* – number of observations, *i* – discrete time.

The aircraft state prediction problem is to find the its future state x_{i+1}, knowing only some past values of states and controls, without any priori information on the aircraft model parameters *A*, *B*.

To solve this problem, let's write the equation (1) in a block-matrix form

$$\begin{bmatrix} X_i & x_{i+1} \end{bmatrix} = A\begin{bmatrix} X_{i-1} & x_i \end{bmatrix} + B\begin{bmatrix} U_{i-1} & u_i \end{bmatrix}, \tag{2}$$

where $X_i = \begin{bmatrix} x_{i-h} & \dots x_i \end{bmatrix}$, $X_{i-1} = \begin{bmatrix} x_{i-h-1} & \dots x_{i-1} \end{bmatrix}$, $U_{i-1} = \begin{bmatrix} u_{i-h-1} & \dots u_{i-1} \end{bmatrix}$, and convert it to the following linear matrix equation

$$\begin{bmatrix} A & B \end{bmatrix}\begin{bmatrix} X_{i-1} & x_i \\ U_{i-1} & u_i \end{bmatrix} = \begin{bmatrix} X_i & x_{i+1} \end{bmatrix}, \tag{3}$$

representing aircraft eigen-dynamics and control efficiency parameters identification problem.

Equation (3) has the form of a right-hand side matrix equation

$$ZC = D,$$

which is solvable if and only if

$$D\overline{C^R} = 0, \tag{4}$$

where $\overline{C^R}$ – the right-hand full rank zero divisor of the matrix *C*, such that $C\overline{C^R} = 0$, which formally describes all the linear dependent columns of matrix *C* [8, 12–15].

According to (4), we always can find such *h*, that the necessary and sufficient solvability condition of (3) has the form of equality

$$\begin{bmatrix} X_i & x_{i+1} \end{bmatrix}\overline{\begin{bmatrix} X_{i-1} & x_i \\ U_{i-1} & u_i \end{bmatrix}}^R = \begin{bmatrix} X_i & x_{i+1} \end{bmatrix}\begin{bmatrix} r_i \\ 1 \end{bmatrix} = 0, \tag{5}$$

where

$$\begin{bmatrix} r_i \\ 1 \end{bmatrix} = \overline{\begin{bmatrix} X_{i-1} & x_i \\ U_{i-1} & u_i \end{bmatrix}}^R \tag{6}$$

is the single-column normalized right-hand zero divisor, such that

$$\begin{bmatrix} X_{i-1} & x_i \\ U_{i-1} & u_i \end{bmatrix}\begin{bmatrix} r_i \\ 1 \end{bmatrix} = 0. \tag{7}$$

So, from (5) we can find a simple nonparametric aircraft state prediction algorithm

$$x_{i+1} = -X_i r_i, \tag{8}$$

allowing obtaining the future aircraft state x_{i+1} based only on some *h* past measurements without a direct solution of identification equation (3) and any information on the model parameters *A* and *B*.

Note that the condition (5) is met for all discrete times, so to predict the next aircraft states we need iterating (7) and (8) for all future steps.

3. Nonparametric aircraft state prediction examples

Consider a simple aircraft roll motion model

$$\begin{cases} \dot{\omega}(t) = k_\omega \omega(t) + k_\delta \delta(t), \\ \dot{\gamma}(t) = \omega(t), \end{cases} \quad (9)$$

where t – continuous time, ω – angular rate of roll, γ – roll, δ – deflection angles of ailerons , k_ω – natural damping coefficient, k_δ – efficiency of ailerons.

With the zero initial conditions $\omega(t)=0$, $\dot{\omega}(t)=0$, $\gamma(t)=0$, $\dot{\gamma}(t)=0$, constant control $\delta(t)=1$, and model parameters $k_\omega=-2$, $k_\delta=2$ we can find the exact solutions of the equations (9)

$$\begin{cases} \omega(t) = 1 - e^{-2t}, \\ \gamma(t) = t + \dfrac{e^{-2t}-1}{2}. \end{cases} \quad (10)$$

According to (10) the first five states with $\Delta t = 0.1$ sec discretization interval are presented in table 1.

Table 1. The exact solutions of aircraft model differential equations.

i	0	1	2	3	4
t	0	0.1	0.2	0.3	0.4
$\omega(t)$	0	0.1813	0.3297	0.4512	0.5507
$\gamma(t)$	0	0.0094	0.0352	0.0744	0.1247
$\delta(t)$	1	1	1	1	1

Let we know only the first four columns in table 1 and we need to predict the fifth one using the proposed method. Then, according to (2) we have the following block matrix of states and controls

$$\begin{bmatrix} X_{i-1} & x_i \\ U_{i-1} & u_i \end{bmatrix} = \begin{bmatrix} 0 & 0.1813 & 0.3297 & 0.4512 \\ 0 & 0.0094 & 0.0352 & 0.0744 \\ 1 & 1 & 1 & 1 \end{bmatrix}. \quad (11)$$

It's easy to check that for the following right-hand single-column zero divisor of matrix (11)

$$\overline{\begin{bmatrix} X_{i-1} & x_i \\ U_{i-1} & u_i \end{bmatrix}}^R = \begin{bmatrix} r_i \\ 1 \end{bmatrix} = \begin{bmatrix} -0.8170 \\ 2.6340 \\ -2.8166 \\ 1 \end{bmatrix}$$

the null space equity (5) is always fulfilled:

$$\begin{bmatrix} X_{i-1} & x_i \\ U_{i-1} & u_i \end{bmatrix} \overline{\begin{bmatrix} X_{i-1} & x_i \\ U_{i-1} & u_i \end{bmatrix}}^R = \begin{bmatrix} X_{i-1} & x_i \\ U_{i-1} & u_i \end{bmatrix} \begin{bmatrix} r_i \\ 1 \end{bmatrix} = \begin{bmatrix} 0 & 0.1813 & 0.3297 & 0.4512 \\ 0 & 0.0094 & 0.0352 & 0.0744 \\ 1 & 1 & 1 & 1 \end{bmatrix} \times \begin{bmatrix} -0.8170 \\ 2.6340 \\ -2.8166 \\ 1 \end{bmatrix} = \begin{bmatrix} 0 \\ 0 \\ 0 \end{bmatrix}.$$

Then, using the expression (8) we can find the predicted aircraft state

$$\hat{x}_{i+1} = \begin{bmatrix} \hat{\omega}_{i+1} \\ \hat{\gamma}_{i+1} \end{bmatrix} = -X_i r_i = -\begin{bmatrix} 0.1813 & 0.3297 & 0.4512 \\ 0.0094 & 0.0352 & 0.0744 \end{bmatrix} \times \begin{bmatrix} -0.8170 \\ 2.6340 \\ -2.8166 \end{bmatrix} = \begin{bmatrix} 0.5507 \\ 0.1245 \end{bmatrix},$$

that entirely congruent with the exact values in last column of table 1.

Consider now the approximate numerical aircraft model (9) solutions using first-order Euler method for solving ordinary differential equations. Then (9) with the same parameters and initial conditions has the following discrete time-invariant form

IOP Conf. Series: Materials Science and Engineering **476** (2019) 012003 doi:10.1088/1757-899X/476/1/012003

$$\begin{cases} \dfrac{\omega_{i+1} - \omega_i}{\Delta t} = k_\omega \omega_i + k_\delta \delta_i, \\ \dfrac{\gamma_{i+1} - \gamma_i}{\Delta t} = \omega_i. \end{cases} \tag{12}$$

Rewriting (12)

$$\begin{cases} \omega_{i+1} = (1 + k_\omega \Delta t)\omega_i + (k_\delta \Delta t)\delta_i, \\ \gamma_{i+1} = (\Delta t)\omega_i + \gamma_i, \end{cases}, \tag{13}$$

brings it to the matrix form (1), where

$$x = \begin{bmatrix} \omega \\ \gamma \end{bmatrix},\ u = \delta,\ A = \begin{bmatrix} 1 + k_\omega \Delta t & 0 \\ \Delta t & 1 \end{bmatrix} = \begin{bmatrix} 0.8 & 0 \\ 0.1 & 1 \end{bmatrix},\ B = \begin{bmatrix} k_\delta \Delta t \\ 0 \end{bmatrix} = \begin{bmatrix} 0.2 \\ 0 \end{bmatrix}.$$

Let the ailerons control strategy be a little bit more complex than in the first example as shown in table 2.

Table 2. Aircraft control values.

i	0	1	2	3	4	5	6	7…20
t_i	0	0.1	0.2	0.3	0.4	0.5	0.6	0.7…2
u_i	1	1	0	0	0.5	0.5	0	0…0

Simultaneous to simulation (13) we also iterate our prediction algorithm (7)–(8) with sample width $h = 4$. The first some simulated ω_i, γ_i and predicted $\hat{\omega}_i$, $\hat{\gamma}_i$ states are shown in table 3, the rest can be seen from figure 1.

Table 3. Aircraft simulated and predicted states values

i	0	1	2	3	4	5	6	7	8	9
t_i	0	0.1	0.2	0.3	0.4	0.5	0.6	0.7	0.8	0.9
δ_i	1	1	0	0	0.5	0.5	0	0	0	0
ω_i	0,00	0,20	0,36	0,29	0,23	0,28	0,33	0,26	0,21	0,17
γ_i	0,00	0,00	0,02	0,06	0,08	0,11	0,14	0,17	0,20	0,22
$\hat{\omega}_i$	-	-	-	-	0,23	0,28	0,33	0,26	0,21	0,17
$\hat{\gamma}_i$	-	-	-	-	0,08	0,11	0,14	0,17	0,20	0,22

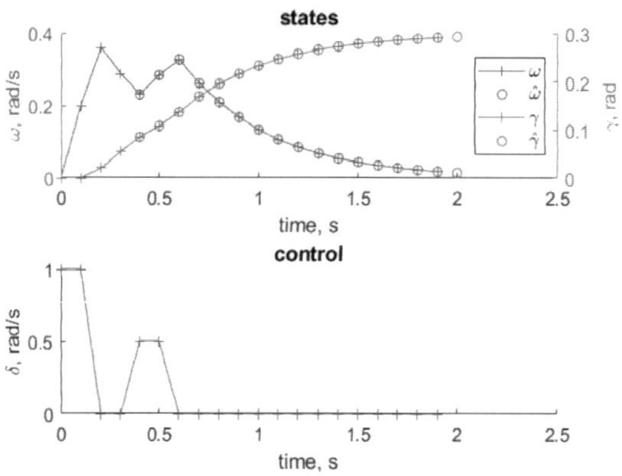

Figure 1. Aircraft control, simulated and predicted states values.

It can be seen, that proposed nonparametric prediction algorithm needs only four simulations steps or 0.4 seconds for tuning in this example. As soon as it collect enough data for equation (7) nontrivial solution existence it start to predict future aircraft states without using information of A and B matrices, just based only on h past measurements and formula (8).

Acknowledgments
This work was supported by Russian Foundation for Basic Research (RFBR), projects 17-08-01445a, 17-01-00251a, 18-08-00453a.

References

[1] Kuznetsov A, Shevchenko A and Solonnikov J 2013 *IFAC Proceedings Volumes* **46**(19) 183–187.
[2] Kim Y, Park J and Bang H 2018 *Navigation* **65**(2) 157–167
[3] Linares R, Vittaldev V and Godinez H C 2018 *Handbook of Dynamic Data Driven Applications Systems* (Springer)
[4] Yu X and Zhang Y 2015 *Progress in Aerospace Sciences* **74** 152–166
[5] Korsun O N, Stulovskii A V, Balyk O A and Zolotaeva M V 2018 *IOP Conference Series: Materials Science and Engineering* **312** 012015
[6] Korsun O N, Stulovskii A V, Ovcharenko V N and Kanyshev A V 2018 *Journal of Computer and Systems Sciences International* **57**(3) 374–389
[7] Korsun O N, Om M H, Latt K Z and Stulovskii A V 2017 *Procedia Computer Science* 67–74
[8] Zybin E Yu 2015 *Izvestiya SFedU. Engineering Sciences* **4**(166) 160–170 (in Russian)
[9] Wang Z, Liang M and Delahaye D 2018 *Transportation Research Part C: Emerging Technologies* **95** 280–294
[10] Hrastovec M and Solina F 2016 *Transportation Research Part C: Emerging Technologies* **73** 167–182
[11] Karlaftis M G and Vlahogianni E I 2011 *Transportation Research Part C: Emerging Technologies* **19** (3) 387–399
[12] Zybin E Y and Kos'yanchuk V V 2016 *Journal of Computer and Systems Sciences International* **55**(4) 546–557
[13] Zybin E, Kosyanchuk V and Karpenko S 2017 *MATEC Web of Conferences* **99** 03011
[14] Karpenko S S, Zybin E Yu and Kosyanchuk V V 2018 *IOP Conference Series: Materials Science and Engineering* **312** 012010
[15] Terent'yev M N, Karpenko S S, Zybin E Yu and Kosyanchuk V V 2018 *IOP Conference Series: Materials Science and Engineering* **312** 012025

IOP Conf. Series: Materials Science and Engineering **476** (2019) 012004 doi:10.1088/1757-899X/476/1/012004

Study of the Impact of Unique Technological Competencies on the Economic Growth of Large Enterprises and High-tech Industries

A A Chursin[1], V V Strenalyuk[1,2] and S A Agaptsov [3]

[1] Peoples' Friendship University of Russia, Moscow 117198, Russia
[2] State Corporation «Rostec», Moscow 119048, Russia
[3] Auditor of the Chamber of Accounts of Russian Federation Moscow, 119435, Russia

E-mail: achursin2008@yandex.ru

Abstract. The article suggests a tool for assessing the impact of unique technological competences on the economic growth of large entreprises, companies and hi-tech industires. The main research method is a classical regression analysis used in economics, based on Cobb-Douglas function. In this work economic-mathematical model of the impact of unique technological competencies on the economic growth of high-tech enterprises and industries, representing a production function, using UTC is proposed. The significant influence of the competence factor on the economic growth of high-tech enterprises and industries is shown.

1. Introduction

To conduct the objective quantitative assessment of the impact of unique technological competencies (UTC) on the economy of enterprises and industries (for example, aerospace), it is convenient to use the regression analysis based on the growth theory [1-3]. It is possible to build an economic-mathematical model that reflects the mutual influence of the UTC, along with other factors of production, on economic indicators. Various problems of economic development analysis and forecast can be solved, using this model.

The general requirements for economic and mathematical models include:

1) adequacy – compliance of the model with its original;

2) objectivity – compliance of scientific conclusions with real conditions;

3) simplicity – clearness of the model from secondary factors;

4) sensitivity – the ability of the model to respond to changes in the initial parameters;

5) stability – a small perturbation of the initial parameters should correspond to a small change in the solution of the problem;

6) universality – model adaptability.

Economics uses the production function to model economic growth. The production function is the analytical relation that links the variable values of costs (factors, resources) with the value of output.

General view of the production function is the following:

$$Y = Y(X_1, X_2, \ldots, X_i, \ldots, X_n), \quad (1)$$

where Y is an indicator, characterizing production results; X is a factor indicator of the i-th production resource; n is the number of factor indicators.

Production functions are determined by two groups of conditions: mathematical and economic. Mathematically, it is assumed that the production function must be continuous and twice differentiable. The economic assumptions are as follows: if at least one production resource is missed, production is impossible.

The most common model that meets the specified requirements to the production functions is the Cobb-Douglas function:

$$Y = a_0 K^{a_1} L^{a_2}, \tag{2}$$

where $a_0, a_1, a_2 > 0$ are constants, $a_1 + a_2 < 1$ is the condition of diminishing return to production factors; K is the amount of funds either in terms of value or in natural quantity, say, the number of machine units; L is the volume of labor resources, also in terms of value, or in natural quantity – the number of workers, man-days, etc., and, finally, Y is the output in terms of value or in kind.

However, the classic Cobb-Douglas production function does not take into account the influence of scientific and technological progress and human capital, therefore, does not meet the objectives of this study.

The Dutch economist Jan Tinbergen, the Nobel Prize winner in economics, attempted to calculate the rate of economic growth, taking into account the factor of technological progress. He improved the Cobb-Douglas function by introducing an indicator of the rate of technological progress:

$$Y = a_0 K^{a_1} L^{a_2} e^{rt}, \tag{3}$$

where e is the base of the natural logarithm, and the e^{rt} term is the factor of the temporary (t) change in scientific and technological progress. However, it is difficult to apply this function [3].

The authors of the Mankiw-Romer-Weil model included human capital *(H)* as an independent factor of economic growth, which has an endogenous character, and the production function acquired the following form:

$$Y = K^{\alpha} H^{\beta} (AL)^{1-\alpha-\beta}, \tag{4}$$

where α is the coefficient of elasticity of output Y according to the factor of physical capital; β is the coefficient of elasticity of output according to the human capital factor; $1-\alpha-\beta$ is the coefficient of elasticity of production according to the factor of labor;AL is the number of units of effective labor.

However, consideration of human capital in the model is made in monetary terms, according to the assessment of investment in human capital – the cost of training, education, career enhancement, health care, migration, information search, etc. At the same time, there is the lack of the influence of the abilities, capabilities and skills of scientists, engineers and other qualified workers to create monopoly products, unattainable for competitors that form the basis of innovation and scientific and technological progress.

To develop these approaches to the consideration of human capital in the economic and mathematical model of economic growth of high-tech enterprises and industries, we propose to consider the possibility to construct a production function with a neutral, according to Harrod, scientific and technical progress, of the following type:

$$Y = a_0 K^{a_1} L^{a_2} C^{a_3}, \tag{5}$$

Where $C \geq 1$ is the number of unique technological competencies, belonging to key employees of an enterprise (industry, economy).

2. Production function construction

To construct an economic-mathematical model, let us transform the equation

$$Y = a_0 K^{a_1} L^{a_2} C^{a_3} \tag{6}$$

into linear form by taking the logarithm

$$\ln(Y) = \ln(a_0) + a_1 \ln(K) + a_2 \ln(L) + a_3 \ln(C). \quad (7)$$

After substitutions

$$\ln(Y) = Y',$$
$$\ln(a_0) = a_0',$$
$$\ln(K) = K',$$
$$\ln(L) = L',$$
$$\ln(C) = C',$$

we get a linear equation:

$$Y' = a_0' + a_1 K' + a_2 L' + a_3 C'. \quad (8)$$

Next, executing the least squares method, let us define the parameters a_0, a_1, a_2, using the economic data of Rostec State Corporation (Table 1).

Table 1. Statistic data for a production function construction

Period	Y = Sales proceeds over a year, billion rubles	K = Balance cost of non-current assets, billion rubles	L = Average staff number X average monthly earnings, billion rubles	C = Estimate for the number of UTC
2011	430	291	9,82	145
2012	497	382	10,17	140
2013	597	591	10,27	165
2014	807	788	13,47	200
2015	1065	946	15,35	212
2016	1268	862	16,22	244
2017	1466	1196	18,43	236

Source: Reports on the implementation of the Innovation Development Program of Rostec State Corporation for 2011-2017

It is difficult to determine the data of quantitative evaluation C of the number of UTC in the Rostec State Corporation enterprises, due to the fact that the system-based work on the UTC identification and description has just started. As of today, preliminary work has been carried out only in two holdings of the Corporation (Table 2).

Table 2. Correlation between the number of UTC and the number of patents

Holding	The number of patents on inventions, obtained in the last three years (2014-2017)	The number of identified UTC
Holding 1	179	16
Holding 2	320	37

Source: compiled by the author.

When there is the lack of data on the total number of UTC in the Corporation, it can be assumed that the number of UTC is proportional to the number of patents on devices and production methods. The number of patents on inventions, obtained in the last three years before the reporting progressive total should be taken into account, since the introduction of the invention and its economic effect is delayed. While constructing the model, we will assume that the estimate of the number of UTC is 1/10 of the number of patents for three years. This assumption corresponds to the practice of "umbrella" patenting, when approximately 10 patents are issued to protect one basic technology.

Using the method of least squares, we get the following values of model parameters:

$$a_0 = 1.076396232,$$
$$a_1 = 0.204234015,$$
$$a_2 = 1.023199735$$
$$a_3 = 0.507745261$$

3. Checking the adequacy of the production function

In the initial approximation, the adequacy of the constructed mathematical model can be estimated by comparing the graphs of measured and calculated values of the sales revenue Y.

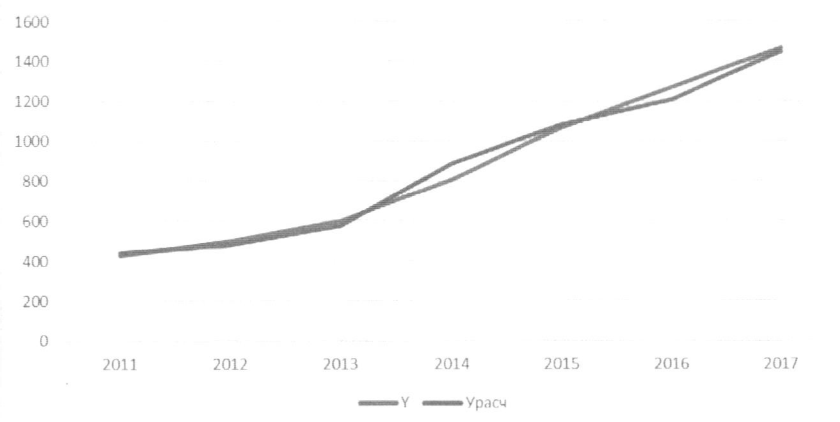

Figure 1. Comparison of statistical and calculated data

The pure mathematical test of the adequacy of the model is carried out, using the Fisher criterion.

Fisher criterion:

$$F_{est} = \frac{\sum (Y_{i\,est} - Y_{avg\,est})^2}{m} \cdot \frac{n - m - 1}{\sum (Y_i - Y_{i\,est})^2} \tag{9}$$

Where $\alpha = 0.99$ is the probability belief, m = 3 is the number of factors, n = 7 is the number of observations.

The mathematical model is considered adequate if

$$F_{est} > F_{table}, \tag{10}$$

where F_{table} can be determined with the help of the F.INV program of MS Excel.

In this case,

$$F_{est} = 76.51887262$$
$$F_{table} = 29.45669513$$

4. Production function analysis

Let us conduct an economic analysis of the results (Table 3).

Table 3. Economic results of the model

Period	μ_K	μ_L	μ_C	ν_K	ν_L	ν_C	γ_{KC}
2011	1,523908	45,193	3,068774	0,311234	46,24147	1,558155	5,006381
2012	1,249563	46,9907	3,421122	0,255203	48,08087	1,737059	6,806566
2013	0,970593	55,86734	3,481345	0,198228	57,16344	1,767636	8,917181
2014	1,124407	65,76124	4,432301	0,229642	67,28688	2,25048	9,79994
2015	1,144714	70,52484	5,109868	0,23379	72,161	2,594511	11,09763
2016	1,400272	74,38259	4,951783	0,285983	76,10824	2,514244	8,791578
2017	1,208185	78,4135	6,134592	0,246753	80,23268	3,11481	12,62321

Source: compiled by the author.

Average resource efficiency:

$$\mu_K = \frac{Y}{K} = \frac{a_0 K^{a_1} L^{a_2} C^{a_3}}{K} = a_0 K^{a_1 - 1} L^{a_2} C^{a_3},$$

$$\mu_L = \frac{Y}{L} = \frac{a_0 K^{a_1} L^{a_2} C^{a_3}}{L} = a_0 K^{a_1} L^{a_2 - 1} C^{a_3},$$

$$\mu_C = \frac{Y}{C} = \frac{a_0 K^{a_1} L^{a_2} C^{a_3}}{C} = a_0 K^{a_1} L^{a_2} C^{a_3 - 1}.$$

Comparing the average resource efficiency μ_L, μ_K и μ_K, we can see that the average payoff from the use of competencies essentially exceeds the average payoff from the use of capital. The average payoff from labor is even higher, which is typical for production functions, but in practice, the increase of labor consumption is possible to a very limited extent.

The average efficiency of capital for 7 years is decreasing, while the average efficiency of labor and competences is growing.

Resource limiting efficiency:

$$\nu_K = \frac{\partial Y}{\partial K} = a_0 a_1 K^{a_1 - 1} L^{a_2} C^{a_3},$$

$$\nu_L = \frac{\partial Y}{\partial L} = a_0 a_2 K^{a_1} L^{a_2 - 1} C^{a_3},$$

$$\nu_C = \frac{\partial Y}{\partial C} = a_0 a_3 K^{a_1} L^{a_2} C^{a_3 - 1}.$$

The resource limiting efficiency $\nu \nu_K, \nu_L$ and ν_C determines the extent to which the output will increase and, consequently, the revenue *Y*, if the corresponding resource (for example, the amount of fixed capital) increases by one. The limiting efficiency of the resource ν_C is more than a sequence higher than the limiting efficiency of the capital ν_K.

Elasticity of output, depending on resource consumption:

$$\delta_K = \frac{\partial Y}{\partial K}\frac{K}{Y} = \frac{a_0 a_1 K^{a_1 - 1} L^{a_2} C^{a_3} K}{a_0 K^{a_1} L^{a_2} C^{a_3}} = a_1,$$

$$\delta_L = \frac{\partial Y}{\partial L}\frac{L}{Y} = \frac{a_0 a_2 K^{a_1} L^{a_2 - 1} C^{a_3} L}{a_0 K^{a_1} L^{a_2} C^{a_3}} = a_2,$$

$$\delta_C = \frac{\partial Y}{\partial C}\frac{C}{Y} = \frac{a_0 a_3 K^{a_1} L^{a_2} C^{a_3 - 1} C}{a_0 K^{a_1} L^{a_2} C^{a_3}} = a_3.$$

The elasticity of resources in the model is constant for all resources. The coefficient $a_3 = 0.507745261$, for example, shows that the increase of the number of used UTC by 1% leads to the revenue increase by 0.51%. In relation to the data of 2017, it means that the addition of one UTC adds $0.51/2.36 \cdot 1466 = 317$ billion rubles to the revenue of the Corporation on average.

Resource replacement rates:

$$\gamma_{KC} = \frac{\nu_C}{\nu_K} = \frac{\partial Y}{\partial C} : \frac{\partial Y}{\partial K} = \frac{a_0 a_3 K^{a_1} L^{a_2} C^{a_3 - 1}}{a_0 a_1 K^{a_1 - 1} L^{a_2} C^{a_3}} = \frac{a_3 K}{a_1 C},$$

$$\gamma_{LC} = \frac{\nu_C}{\nu_L} = \frac{\partial Y}{\partial C} : \frac{\partial Y}{\partial L} = \frac{a_0 a_3 K^{a_1} L^{a_2} C^{a_3 - 1}}{a_0 a_2 K^{a_1} L^{a_2 - 1} C^{a_3}} = \frac{a_3 L}{a_2 C}.$$

The rate of resource replacement shows the proportion of resources interchangeability at the same output.

If we reduce the consumption of resource C by 0.42% (this is equivalent to losing one UTC), we must increase (build up) the consumption of resource K (fixed capital) by 5.36%, in other words, by 64 billion rubles to maintain the output level.

We see the impact of unique technological competences level on economic growth as predicted in [4-7]. The practical ways of competence management are outlined in [8-9] for machine-buliding industry.

5. Conclusion

1. An economic-mathematical model of the impact of unique technological competencies on the economic growth of high-tech enterprises and industries, representing a production function, using UTC as one of the factors is proposed.
2. The model demonstrated the adequateness to the observed data, illustrated by the economic parameters of one of the large high-tech corporations.
3. The significant influence of the competence factor on the economic growth of high-tech enterprises and industries is shown.

Acknowledgments
This paper was financially supported by the Ministry of Education and Science of the Russian Federation on the project No.26.1146.2017/4.6 «Development of mathematical methods to forecast efficiency of using space services in the national economy».

References

[1] Lomkova E N and Epov A A 2005 *Economic-mathematical Models of Production Management. Theoretical* aspects (Volgograd State Technical University).
[2] Vagapova Ya Ya 2007 *Modeling of Economic Growth, Taking into Account Ecological and Social Factors* (Moscow: MAKS Press)
[3] Shalimov S M 2015 Models of Exogenous and Endogenous Economic Growth in the Context of the Development of the Russian Economy *Regional Economy and Management Journal* **4** 8
[4] Chursin A A, Shamin R V and Fedorova L A 2017 The mathematical model of the law on the correlation of unique competencies with the emergence of new consumer markets. *European Research Studies Journal* **3A** 39.
[5] Tyulin A E, Ostrovskaya A A and Chursin A A 2015 *Fundamentals of Innovation Process Management in Knowledge-intensive Industries. (Theory)* (Moscow: Innovative Engineering)

[6] Kashirin A, Semenov A, Ostrovskaya A and Kokuytseva T 2016The Modern Approach to Competencies Management Based on IT Solutions *JIBC-AD - Journal of Internet Banking and Commerce* **01** 1-12

[7] Kashirin A, Semenov A, Ostrovskaya A, Kokuytseva T and Strenaluk V 2016 The Modern Approach to Competence Management and Unique Technological Competences *Quality – access to success* **17** 154 p 105-109

[8] Chursin A, Kashirin A, Strenalyuk V, Semenov A, Ostrovskaya A and Kokuytseva T 2017 The approach to detection and application of the company's technological competences to form a business-model *IOP Conference Series Materials Science and Engineering* **312** 012003

[9] Kashirin A I 2018 Algorithm of Identification and Search for New Market Applications of Unique Technological Competences *European Research Studies Journal* **21(**4) 119-128

IOP Conf. Series: Materials Science and Engineering **476** (2019) 012005 doi:10.1088/1757-899X/476/1/012005

Tool for Assessing the Risks of R&D Projects Implementation in High-tech Enterprises

R A Chursin[1], A V Yudin[1,2], P Yu Grosheva[1,2], P G Filippov[2,3] and E V Butrova[2]

[1] JSC Russian Space Systems, Moscow 111250, Russia
[2] Peoples' Friendship University of Russia, Moscow 117198, Russia
[3] State Corporation «Rostec», Moscow 119048, Russia

E-mail: achursin2008@yandex.ru

Abstract. The article gives a tool for assessing the risks of R&D projects implementation in high-tech enterprises including aerospace industry. The tool is algorithms and calculation formulas that help to make risk assessments at each stage of project implementation. Special attention is attached to the analysis of the most common risks that arise in the course of R&D projects implementation. These risks are divided into three groups: risks that affect the project R&D projects deadlines; risks, affecting the R&D projects financing; risks, associated with the technological characteristics of the projected facility features. This can be fully applied for innovative projects in aerospace industry including development of new materials and composite technologies. The calculation algorithms of the techniques are adapted for computer calculations and the use of statistical data instead of expert estimates contributes to the increase of accuracy and objectivity of calculations

1. Introduction

A high-tech enterprise including aerospace industry may face various risk factors under a modern market economy and an unstable political situation in the world. Such factors can adversely affect the financial condition of enterprises, influence the R&D projects implementation deadlines. For knowledge-intensive industries (including aerospace), to which the enterprises of high-tech industry are fully applied, many risks are connected, first of all, with external factors [1,2]. In the aerospace industry such risks typically include the delay of financing, the changes in construction plan, the negative experimental results in the trial use of new materials and technologies. Also play serious role system risks which are connected with the mistakes and disagreements among partners in consortiums typical for aerospace industry organization.

The diagnosis of probable risk factors for high-tech industry enterprises is one of the most important components of economic analysis [3,4]. Enterprise's management considers its results, while operating the financial policy. Therefore, it is necessary to have techniques that allow to assess possible risks, identify unfavorable factors for the enterprise's activities, further develop a set of anti-risk measures, based on available facilities and resources and calculate the predicted efficiency of these activities (for example, in [5]). Aerospace industry is among the first which started the use of the project management systems as new competences [6] (including software tools) and even was one of the main drivers for PM standards, but the practice showed that the risk assessment tools should be improved.

The sources of the emergence of risk factors for knowledge-intensive enterprises of the aero, rocket and space industry are diverse. For example, the overall economic situation in the country, the unstable political situation in the world, the effect of export sanctions, the activities of competing enterprises abroad. In relation to an enterprise, such risk factors are external, and their management can be difficult due to the complex predictability or insurmountable nature. Other risk factors that are internal in relation to the enterprise, on the contrary, can be well-predicted, their management can be reduced to the timely removal of "pain points" in the internal activities of an enterprise. For example, such factors may be an inexpedient increase of document agreement deadlines within an organization, insufficiency of organization's circulating assets, which are necessary to pay for certain needs, ensuring correct work operation, inadequacy of purchased equipment to the declared technical characteristics. Industry risk can also be connected with the scientific problems when the experimental results do not coincide with the attended metrics and qualities, what is especially typical for new materials.

2. Methods

The process of risk analysis involves the consideration of many external and internal factors with the help of special economic and mathematical methods that allow obtaining estimates of the level of risk and calculate the efficiency of anti-risk measures, on the basis of which risk management will be carried out.

To assess the impact of risk factors on the stage of R&D projects implementation deadlines, let us use the following ratio:

$$R_T = \frac{T_{fact} - T_{plan}}{T_{plan}}, \tag{1}$$

where T_{fact} is the value of the actual time of a project stage implementation, T_{plan} is the planned time of a project stage implementation.

Let us consider the risk of a project cost exceeding the same way:

$$R_P = \frac{P_{fact} - P_{plan}}{P_{plan}}, \tag{2}$$

where P_{fact} is the value of the actual cost of a project stage implementation, P_{plan} is the planned cost of a project stage implementation.

If the project stages (a group of some n project stages) are implemented consistently, then we will consider the following value as a risk assessment, related to a project implementation deadline:

$$R_T = \frac{(T^1_{fact} - T^1_{plan}) + ... + (T^n_{fact} - T^n_{plan})}{T^1_{plan} + T^n_{plan}}, \tag{3}$$

If the project stages (a group of some n project stages) are implemented simultaneously, in parallel and independently of the mutual implementation results, we will consider the following value as a risk assessment, related to a project implementation deadline:

$$R_T = \max_i \frac{(T^i_{fact} - T^i_{plan})}{T^i_{plan}}, \tag{4}$$

where $i = 1, \ldots, n$ corresponds to a project stage, implemented in parallel.

For risk factors that affect the project cost, let us use the following forecast value, regardless of the consecutive or parallel nature of the project stages implementation:

$$R_P = \frac{(P^1_{fact} - P^1_{plan}) + ... + (P^n_{fact} - P^n_{plan})}{P^1_{plan} + P^n_{plan}}. \tag{5}$$

The proposed method for risk assessment can be a part of a more general risk management system that can be implemented in an enterprise or an entire industry. Risk management in the economy is understood as a specific sphere of economic activity that requires deep knowledge of an industry, which is represented by an enterprise, knowledge in the analysis of modern business (an enterprise can implement commercial projects, double-purpose projects, purchase components from third-party suppliers), knowledge of special methods of analysis, assessment and management of risks, as well as in the field of insurance (since insurance and self-insurance are powerful tools of risk management). The main task of an enterprise administration when managing risk is to find an option that would ensure an optimal balance of risk and project objectives achievement (related to income, financial, etc.). The management process is a systematic work of risk analyzing, developing and taking appropriate measures to minimize it. The risk management process generally corresponds to the chart shown in Figure 1. Risk assessment is an integral part of the quantitative risk analysis stage.

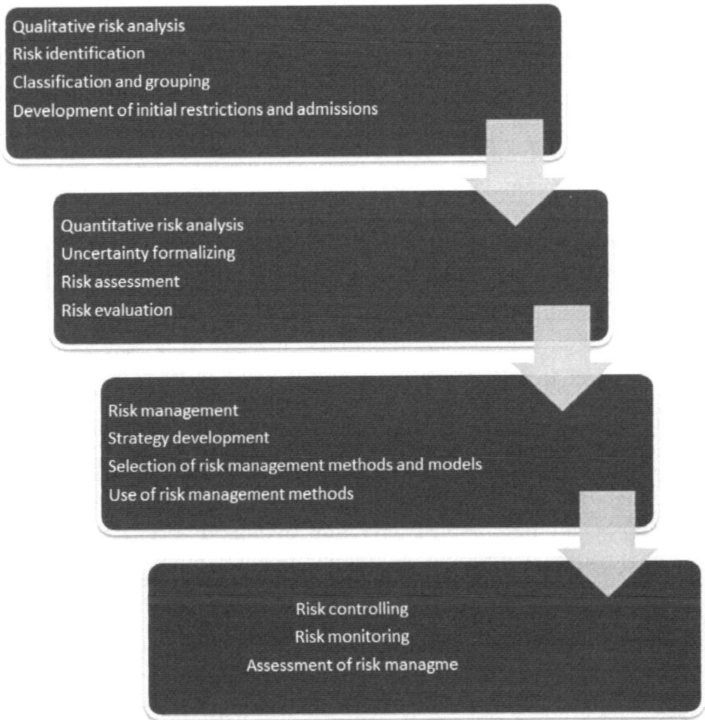

Figure 1. General chart of risk management

Creating a technique for R&D projects risk assessment of Rocket and Space enterprises, we will rely on international risk management standards: ISO 31000: 2009 "Risk management. Principles and guidelines", ISO / IEC 31010: 2009 "Risk management - Risk assessment techniques". These standards stipulate the basic general methods that must be applied when assessing risks.

The initial data for the calculation of the risk assessment algorithm for the R&D projects implementation are the specific risk factors that are most common in the enterprises activities.

The factors, influencing the terms of a project implementation are the following:

- An inexpedient increase in the term of documents coordination both with a customer and intra-company (F_1^1);

- Procurement time increases the project deadline, if there is an increase in the time required to conduct competitive procurement procedures, the time, necessary for the procurement coordination and justification to a customer, organization's internal services or a supplier cannot deliver products on time for whatever reason (F_2^1).

- Co-executor within the organization (another thematic subdivision) cannot deliver his part of the work on time for whatever reason (F_3^1);

- A contractor, engaged at work works cannot deliver his part of the work on time for whatever reason (F_4^1);

- Lack of opportunity to deliver part of the work on time due to insufficient prepayment from a customer (F_5^1);

- Insufficiency of circulating assets of an organization, necessary to pay for various demands, ensuring proper work delivery (F_6^1).

The factors, affecting the project cost are the following:

- Increase in the cost of equipment or materials, necessary for work delivery (F_1^2);

- Increase in the cost of work, performed by a contractor (F_2^2);

- Increase in the cost, paid by an organization from the circulating assets, which affect the ability to deliver work (F_3^2);

- Lack of full prepayment from a customer (F_4^2);

- The necessity for extra costs for work delivery due to newly discovered structural and technological features of the projected facility (F_5^2).

The factors, related to the features of technological characteristics of the projected facility are the following:

- Lack of purchase of necessary equipment due to export sanctions, lack of available sales quota in a supplier (F_1^3);

- In the course of the work it turned out that the appropriate implementation of the technological characteristics of a projected facility will require financial expenses, exceeding the expected useful result (F_2^3);

- Absence of modelling in the life cycle of a product (a customer does not usually include it in the contract performance sheet), that is why many constructive and technical problems can be detected only at the adjustment stage after the first prototypes have been produced (F_3^3);

- Non-compliance of purchased equipment to the declared specifications (F_4^3).

3. Algorithms for assessing the risks of R&D projects implementation in high-tech enterprises

1. Algorithm for assessing risks that affect a project deadline

To assess the risks that affect the project deadline, we will use the method, recommended by ISO/IEC 31010-2011 "Risk management. Methods of risk assessment", which is related to the construction of a matrix of consequences and probabilities. The specific character of the risk technique, considered in this section is that an increase in the project implementation time is the only considered consequence of these factors.

Supposing, the considered stage of a project requires the estimation of the possible excess of the project implementation terms. In other words, it is necessary to determine the T_{fact} value of a project cost. Let the risk management group of an enterprise identify possible risk factors, which may have a negative impact on the R&D projects implementation at this stage. Let us first consider the case when the distinguished risk factor is the only one. In this case, the value T_{fact} of time for a project stage implementation can be determined by the following formula:

IOP Conf. Series: Materials Science and Engineering **476** (2019) 012005 doi:10.1088/1757-899X/476/1/012005

$$T_{fact} = T_{plan} + \Delta T;$$
$$\Delta T = k \cdot T_{plan};$$
$$k > 0,$$

where k is a specified coefficient.

The value of the k coefficient can be learnt in two ways: by using statistical data or by an expert method. Let us consider both of these methods.

The method, based on statistical data resides in analyzing the previous experience of projects implementation, marked by the negative impact of certain risk factors. Basing on the statistical data for the considered risk, we can comprise series that can help to determine the mathematical expectation of exceeding the project implementation time:

Let us consider the average value of the excess

$$\Delta T_{cp} = \frac{1}{n}\sum_{i=1}^{n} t_i \quad (6)$$

as a probable estimate of the excess of a project stage implementation time, constructed according to statistical data.

Thus, the following estimate is correct:

$$E(T_{fact}) = T_{plan} + E(T_{cp}) . \quad (7)$$

Another method is to obtain an expert estimate of a project stage implementation time excess by filling the pairwise comparison matrix. As objects for comparison, the expert is offered to pairwise assess the possible excess of project stage implementation time, expressed as a percentage of the planned time. Herewith, the eigenvalue vector of such matrices will be a priority vector.

To compare two objects, using the pairwise comparison method, the following rating scale, offered by T. Saati is often used. So, let there be given 2 objects A and B (exceeding time) and a comparison criterion. We will fill in the integer values of estimates from 1 to 9 according to the following rule:

- Estimate 1 is filled in if object A and object B are equally preferable.
- Estimate 3 is filled in if object A is slightly preferable than object B. Consequently, the estimate 1/3 corresponds to the situation when A is slightly less preferable than B.
- Estimate 5 is filled in if object A is significantly more preferable than object B. Consequently, the estimate 1/5 corresponds to the situation when A is significantly less preferable than B.
- Estimate 7 is filled in if object A is obviously more preferable than object B in relation to the comparison criterion. Consequently, the estimate 1/7 corresponds to the situation when A is obviously less preferable than B.
- Estimate 9 is filled in if the significance of the object A is absolutely more preferable than the object B. Consequently, the estimate 1/9 corresponds to the situation when A is absolutely less preferable than B.
- Estimates 2, 4, 6, 8 (1/2, 1/4, 1/6, 1/8) are used to facilitate trade-offs between arguments that differ slightly between the two basic numbers.

Let us give an example of the matrix of pairwise comparisons of a project stage implementation time exceeding as a result of the negative impact of a risk factor F_1^1 (Table 1).

Table 1. Example of the matrix of pairwise comparisons

5%	10%	15%
1	1/3	1/2
3	1	2
2	1/2	1
2	1/2	1

The vector of priorities of this matrix of pairwise comparisons is (0.122, 0.424, 0.227, 0,227).

In order to obtain k, corresponding to the considered situation, it is necessary to obtain the sum of the products of the considered relative exceedings for the corresponding component of the priority vector. For our example, the result is k= 5%·0.122+10%·0.424+15%·0.227+20%·0.227=12.795%.

Let T_{plan} =*100*, then, using the formula, we obtain T_{fact} = 100+0.128·100=112.8

If the negative effect of several risk factors, affecting the project implementation time are predicted, the formula for obtaining the final value of the R&D project stage implementation time excess takes the following form in the case of simultaneous influence of factors:

$$T_{fact} = \max_i (T_{plan} + \Delta T_i) . \tag{8}$$

where i corresponds to the risk factor F_i^1.

2. Risks, related to an increase in the project cost.

For risks, related to the change in the project cost, the disquisitions on their evaluation are similar to the case when risk factors affect the R&D projects deadlines.

In this case, the P_{fact} value of time for a project stage implementation can be determined according to the following formula:

$$\begin{gathered} P_{fact} = P_{plan} + \Delta P; \\ \Delta P = k \cdot P_{plan}; \\ k > 0, \end{gathered} \tag{9}$$

where k is a certain coefficient.

To assess such risks, it is also advisable to use methods, based on statistical data or methods of expert evaluation.

If the negative effect of several risk factors, affecting the project cost are predicted, the formula for obtaining the final value of the R&D project stage implementation time excess takes the following form in the case of influence of factors on the same project stage:

$$P_{fact} = \sum_{i=1}^{n} (P_{plan} + \Delta P_i), \tag{10}$$

where i corresponds to the risk factor F_i^2).

3. Risks, related to the technological characteristics of the projected facility features.

The features of the risk factors of this group is the ability to manage them at the prediction stage. Let us hereafter assume that the negative impact of these factors is the increase of a project cost. In this case, it is advisable to predict the impact of risk factors, taking into account the possible financial costs on anti-risk measures. Using simulation modeling, it is possible to calculate the optimal amount of money, spent on the anti-risk measures. To do this, we can use the algorithms, offered in the previous subsection.

In this case, the economic effect B from the measures implementation will be equal to the amount of savings:

$$B = (P_{fact} - P_{plan}) - D, \tag{11}$$

where D is the sum of expenses on anti-risk measures.

4. Conclusion

The work offers the technique of the estimation of R&D projects risks in hi-tech enterprises including aerospace industry. The technique contains algorithms and calculation formulas, used to estimate risks at each stage of the R&D project implementation. The creation and implementation of an efficient R&D management and control system is possible due to the improving of methods for assessing the projects investment attractiveness and risks, as well as the development of a corporate R&D management system in the rocket and space industry.

Acknowledgments
The reported study was funded by RFBR according to the research project № 17-02-00658 «Elaboration of mechanisms for effective management of competitiveness of Russian companies in the context of import substitution policies and modern global trends of the political, economic, scientific and technological development».

References
[1] Kashirin A, Semenov A, Ostrovskaya A and Kokuytseva T 2016 The Modern Approach to Competencies Management Based on IT Solutions *JIBC-AD - Journal of Internet Banking and Commerce* **01** 1-12
[2] Kashirin A, Semenov A, Ostrovskaya A, Kokuytseva T and Strenaluk V 2016 The Modern Approach to Competence Management and Unique Technological Competences *Quality – access to success* **17**(154) 105-109
[3] Chursin A, Semenov A and Danilchankau A 2016 Analysis of Innovation Development in the Economy with Exhaustible Resource Sector by First Order Dynamical Systems Application *Nonlinear phenomena in complex systems* **19**(3) 254-270
[4] Chursin A, Kashirin A, Strenalyuk V, Semenov A, Ostrovskaya A and Kokuytseva T 2017 The approach to detection and application of the company's technological competences to form a business-model *IOP Conference Series Materials Science and Engineering* **312** 012003
[5] Shamin R, Chursin A, Kokuytseva T, Ostrovskaya A and Semenov A 2018 Modeling of Growth-Collapse Processes and Their Applications to Pricing Management *International Journal of Pure and Applied Mathematics* **118**(18) 3741-3746
[6] Kashirin A I 2018 Algorithm of Identification and Search for New Market Applications of Unique Technological Competences *European Research Studies Journal* **21**(4) 119-128

IOP Conf. Series: Materials Science and Engineering **476** (2019) 012006 doi:10.1088/1757-899X/476/1/012006

Influence of Atmospheric Parameters on Aircraft Characteristics

A I Daneko[1], A N Korochkov[1], A Yu Korolev[1], M N Pravidlo[1] and A A Yatsko[2]

[1] Moscow Aviation Institute (National Research University), Moscow, 125319, Russia
[2]Bauman Moscow State Technical University, Moscow, 105005, Russia

*E-mail: a.y.korolyov@yandex.ru

Abstract. In this paper, a comparison of the standard atmosphere from ISO 4401-81 and several empirical models of the atmosphere obtained from sounding data is carried out. The main purpose of the work is to establish the influence of parameters of the atmosphere on the aerodynamic characteristics of an aircraft.

1. Introduction

Many developers use the standard atmosphere (SA) ISO 4401-81 model in the design of aircraft. According to SA at the increase of altitude H air temperature t_a first linearly decreases from 15 ^{0}C to -56,5 ^{0}C (at H=0...11 km), then has a constant value -56,5 ^{0}C (at H=11...20 km) and then increases from -56,5 ^{0}C to -2,5^{0}C (at H=20...47,4 km). The other parameters of the atmosphere in) ISO 4401-81 are determined using equations from a mandatory application. So, the parameters of the atmosphere are functions of altitude.

However, the parameters in real atmosphere also depend on weather conditions established at the moment in an area. It means that SA gives only an average estimate of atmosphere parameters. This can lead to errors in the calculation of aerodynamic and flight characteristics of an aircraft for weather conditions different from the conditions of SA.

More accurate data on parameters of the atmosphere corresponding to a specified time and specified place, may be obtained with sounding of atmosphere, which are the mart of meteorological observations. The use of such data makes it possible to improve the accuracy of calculations in problems that require the assessment of aerodynamic and flight characteristics of an aircraft, as well as in aircraft parameter identification [1], aircraft control [2], onboard measurement errors estimation [3].

2. Comparison of parameters of standard atmosphere and empirical models of the atmosphere

We will carry out a comparative analysis of SA and several empirical models of the atmosphere (EMA) in the altitude range from ground level to H=30 km. To do this, we take the atmosphere model from ISO 4401-81 and four EMA defined on a basis of $t_a(H)$. The dependences $t_a(H)$ are taken from [4] and correspond to the data of serological atmospheric sounding carried out in January and July 2017 at various meteorological observation stations. Information about these stations, as well as the dates and times of the sounding, is presented in table 1.

Table 1. Information of meteorological observation stations.

Number of meteorological observation station	Nearest town	State	Date and time of sounding
24266	Verkhoyansk	Russia	30 January 2017, 12.00 UTC
34882	Astrakhan	Russia	30 January 2017, 12.00 UTC
34882	Astrakhan	Russia	24 July 2017, 12.00 UTC
40417	Dammam	Saudi Arabia	24 July 2017, 12.00 UTC

First, let's compare the dependence of the air temperature ta on the altitude. Figure 1 shows that all considered EMA have significant differences from the SA in nature of changes of air temperature in height. Indeed, both the rate of air temperature change in an troposphere and stratosphere and the location of the boundaries between these atmospheric layers are different from the SA.

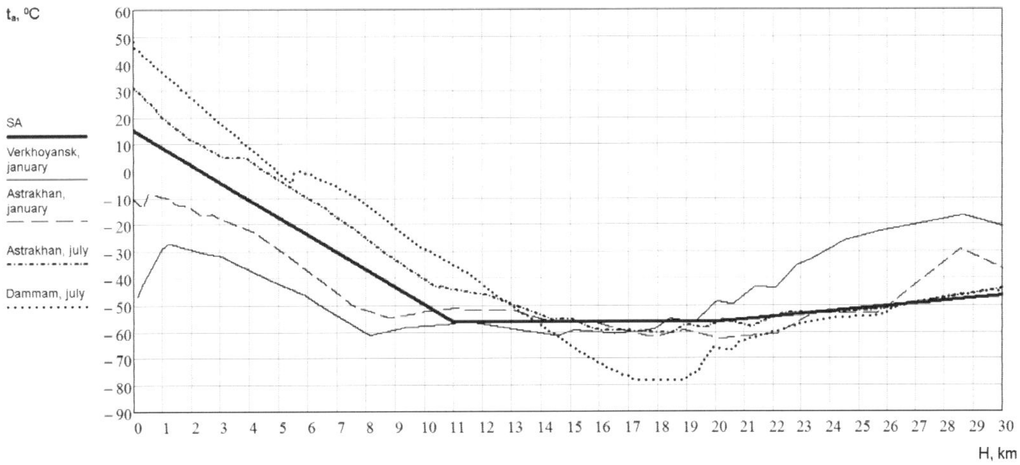

Figure 1. The air temperature in SA and EMA

If $t_a(H)$ is known, we can determine a dependence of the sonic speed and a dynamic air viscosity coefficient μ_a as a function of height

$$a = 20{,}0468 \cdot \sqrt{t_a + 273.2} \tag{1}$$

$$\mu_a = \frac{1{,}458 \cdot 10^{-6}}{t_a + 383{,}6} \cdot (t_a + 273.2)^{\frac{3}{2}} \tag{2}$$

Formulas (1) and (2) are taken from the mandatory Appendix to ISO 4401-81. These formulae for considered height ranges are valid not only for SA, but also for any other model of the Earth's atmosphere. This is due to the fact that the proportional ratio of gases in the air is constant throughout the near-earth space up to H=90-95 km. The dependence $a(H)$ determines the Mach number at a known velocity of the incoming air flow, and $\mu_a(H)$ determines the Reynolds number. The values of the Mach and Reynolds numbers are used in determining aerodynamic characteristics of aircraft.

Let us determine atmospheric pressure p_a and the density of air ρ_a as functions of height H for SA and the four models of EMA described above. To do this, we solve a system of two equations: the equations of state of an ideal gas (3) and the equations of atmospheric statics (4).

$$p_a = 287{,}053 \cdot \rho_a \cdot (t_a + 273{,}2) \tag{3}$$

$$\frac{dp_a}{dH} = -\rho_a \cdot g,\ H=H_0,\ p_a=p_{a0}\ , \tag{4}$$

where H_0 – height of ground level relative to sea level; p_{a0} – atmospheric pressure at ground level, g – acceleration of gravity, defined by the formula (5)

$$g = 9{,}7803(1 + 5{,}302 \cdot 10^{-3} \cdot sin^2(\Psi_E)) - 3{,}086 \cdot 10^{-3} \cdot H, \qquad (5)$$

where Ψ_E – geographic latitude, rad , H – height, km.

Equations (3) and (4) are also taken from the mandatory Appendix to ISO 4401-81. Similar to equations (1) and (2), these equations are applicable in the considered altitude range not only for SA, but also for any other model of the earth's atmosphere, including the considered EMA.

The values of Ψ_E, H_0 and p_{a0} are presented in table 2.

Table 2. Initial conditions for solving the system of equations (3), (4)

Atmospheric model	Town	Month	Ψ_E, degree	H_0, km	p_{a0}, Pa
SA	---	---	45,3	0	101325
EMA	Verkhoyansk	January	67,6	0,138	98800
EMA	Astrakhan	January	46,3	-0,022	103100
EMA	Astrakhan	July	46,3	-0,022	101200
EMA	Dammam	July	26,4	0,012	99400

Having solved the system of equations (3) and (4), we compare the differencies in $p_a(H)$ and $\rho_a(H)$ between SA and the considered EMA. To do this, imagine the values of p_a and ρ_a in EMA as a percentage deviation with respect to p_a and ρ_a in SA. These deviations $\delta p_a(H)$ and $\delta \rho_a(H)$ are shown in figure 2 and 3.

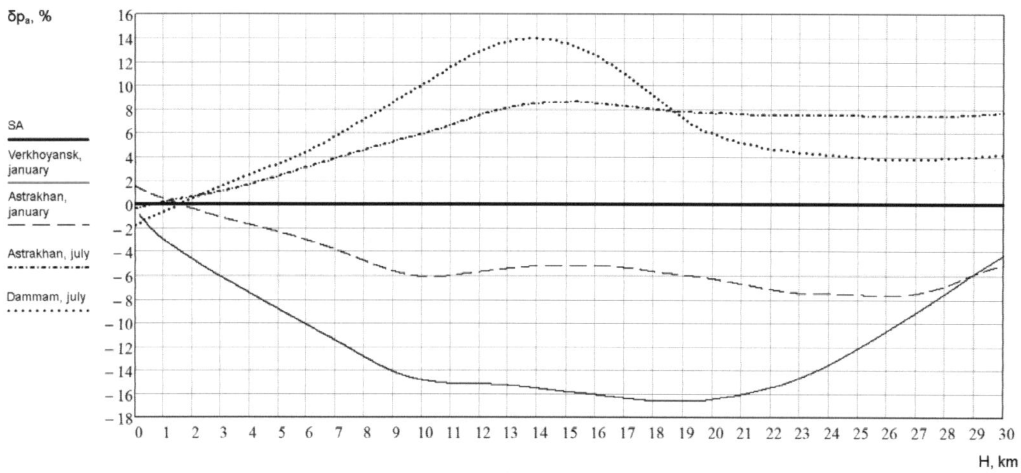

Figure 2. Atmospheric pressure in EMA relative to atmospheric pressure in SA

As shown in figure 2, atmospheric pressure in the EMA is markedly different from p_a in the SA. In this case, the highest value $|\delta p_a|$ are obtained at medium and high altitudes. The decrease in p_a relative to SA is typical for January EMA, and the increase – for July. Function $p_a(H)$ is used in the calculation of the characteristics of the aerial vehicle as a value that determines the altitude correction to the thrust of the engine.

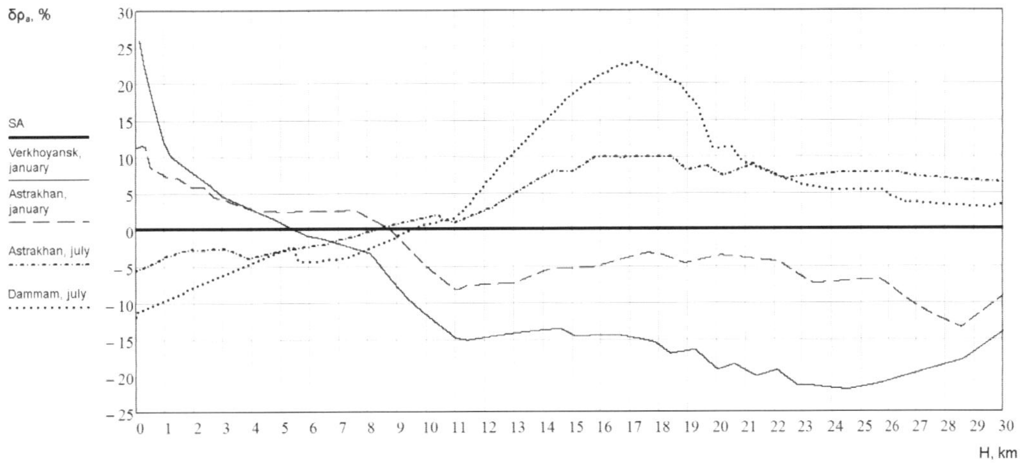

Figure 3. Density of air in the EMA relative to the density of air in the SA

For figure 3 are also visible differences between the density of the air ρ_a in SA and EMA. At low altitudes, the highest ρ_a values are reached in the EMA for Verkhoyansk, and the lowest-in the EMA for the city of Dammam. At large H the most ρ_a is in the city of Dammam and the city of Astrakhan in July and the lowest over the city of Verkhoyansk. The $\rho_a(H)$ dependence is used to determine the dynamic pressure and Reynolds number.

Based on the comparative analysis of SA and EMA, we conclude that the considered models of the atmosphere have significant differences. As it turned out, for a range of heights from ground level up to $H=30$ km, the difference in speed of sound between the SA and EMA is in the range -11...7%, in dynamic coefficient of viscosity is in the range of -17...11%, atmospheric static pressure – in the range -17...14% and the air density – in the range of -22...26%. The greatest deviations from the standard p_a values are observed in EMA, appropriate to the city of Verkhoyansk and the city of Dammam.

3. The algorithm for calculating the aerodynamic forces acting on the aircraft in flight.

When considering the flight of an aircraft in the speed coordinate system vector of the total aerodynamic force can be divided into two components: the drag force of the aircraft Q_a (directed along the velocity vector of the aircraft) and the lift of the aircraft Y_a (directed perpendicular to the velocity vector of the aircraft). The values of Q_a and Y_a are determined according to the formulas (6) and (7). [5]

$$Q_a = \left(c_{xp}(M) + c_{xf}(M, Re) + c_{xaibal}(\alpha, M)\right) \cdot q \cdot S_m\,, \tag{6}$$

$$Y_a = c_{yaбал}(\alpha, M) \cdot q \cdot S_m\,, \tag{7}$$

where c_{xp} – resistance pressure coefficient of aircraft, c_{xf} – resistance friction coefficient of an aircraft, c_{xaibal} – inductive resistance coefficient of an aircraft, M – Mach number (determined by the formula (8)), Re – Reynolds number (determined by the formula (9)), α –angle of attack, q – dynamic pressure (determined by the formula (10)), S_m – characteristic surface of aircraft, c_{yabal} – lift coefficient of the aircraft.

$$M = \frac{V}{a}\,, \tag{8}$$

$$Re = \frac{V \cdot \rho_a}{\mu_a}, \tag{9}$$

$$q = \frac{\rho_a \cdot V^2}{2}, \tag{10}$$

where V – air speed of the aircraft.

From formulas (6) - (10) it may be seen that at the same values of the parameters V, H and α in different models of the atmosphere aerodynamic forces of the aircraft are be different. These differences are determined by differences in the values of the aerodynamic coefficients and the dynamic pressure q, which depends on the air density ρ_a. The coefficients of c_{xf} vary between different models of the atmosphere due to differences in the numbers of M and Re and the coefficients c_{xp}, c_{xr}, c_{xaibal} and c_{yabal} – due to differences in M. Because of the difference in aerodynamic forces of aircraft for different models of the atmosphere, it follows that flight characteristics of aircraft will also depend on the model of the atmosphere. As shown by the study of the flight characteristics hypothetical aircraft, conducted for the SA and four previously considered EMA, the values of the aircraft maximum range D_{max} between the SA and the EMA are different. Thus, when considering the flight of a hypothetical aircraft in January EMA the values of D_{max} at low altitude H=0.5 km are smaller than in SA up to -6%, and at H=11 km – greater more than to 14%.

When considering the flight of a hypothetical aircraft in July EMA, the opposite situation is observed: at low altitudes D_{max} is greater than in SA up to 6% , and at high altitudes – smaller up to -9% at H=15 km.

Therefore, when solving problems that require the precision in calculating the aircraft aerodynamic and flight characteristics, the use of EMA is more preferable than the SA of ISO 4401-81.

4. Conclusion

As a result of comparison of SA and four EMA it is established that the difference between parameters of the atmosphere in SA and considered EMA is in the range from -22% to 26%. Considering the model of aerodynamic forces it was found that in different models of the atmosphere aerodynamic characteristics of an aircraft are different from each other. For a hypothetical aircraft, the values of the maximum ranges D_{max} for SA and four EMA show the difference in the range from -9 to 14%.

References

[1] Korsun O N, Stulovskii A V, Ovcharenko V N and Kanyshev A V 2018 *Journal of computer and systems sciences international* **57**(3) 374-389
[2] Zybin E Y and Kos'yanchuk V V 2016 *Journal of Computer and Systems Sciences International* **55**(4) 546–557
[3] Pushkov S G, Lovitskii L L and Korsun O N 2018 *Measurement Techniques* **57**(3) 140-147
[4] *Atmospheric soundings* (Univ. of Wyoming) http://weather.uwyo.edu/upperair/sounding.html
[5] Lebedev A, Chernobrovkin L 1962 *Dynamics of flight*, pp 121-128 (In Russian)

IOP Conf. Series: Materials Science and Engineering **476** (2019) 012007 doi:10.1088/1757-899X/476/1/012007

An Approach to the Feature Selection in the Objects Detection Problem in Images

L A Demidova[1,2], **M M Egin**[1] **and R V Tishkin**[3]

[1]Ryazan State Radio Engineering University, Ryazan, Russia
[2]Moscow Technological Institute, Moscow, Russia
[3]Limited Liability Company "Areal-98", Moscow, Russia

E-mail: demidova.liliya@gmail.com

Abstract. The objects detection problem in images has been considered. It is known that this problem can be solved using the SVM classifier development with application of the descriptors of the histograms of oriented gradients for creation of the training and test sets, but this technics deals with the significant time expenses for the formation of the histograms and the development of the SVM classifier. The application of the special tool known as the feature selection using the neighborhood component analysis algorithm for classification to determine the most informative features of objects has been investigated. It is shown that this approach allow to reduce the time expenses on solving the objects detection problem while maintaining acceptable values of the classification quality indicators. The experimental results on solving the objects detection problem have been presented.

1. Introduction

The information technologies development has led to the emergence of the digital types of information, which made it possible to obtain the data using the automated systems. To realize this possibility, a large number of various algorithms, which allow analyzing the video data, can be used. The video data analysis is associated with the concept of the "video analytics". Video analytics is a technology that uses the computer vision techniques to solve various video data analysis problems.

The objects detection problem using the images or video data is one of the important problems of the computer vision. The purpose of detection is to determine whether there is a desired object on the incoming image.

Algorithms that solve this problem are used in the development of the modern interfaces for the interaction of systems and humans. Such algorithms are, for example, used in the robotics, in the tracking systems, in the security systems, etc. The problem of detection is reduced to finding the objects of a predetermined class in the image. In this case, when selecting objects in the image, the desired objects can be considered as the sets of pixels corresponding to them, or as the conditional rectangles bordering these objects. In this paper, the selection of the desired objects in the rectangles will be used, and as a toolkit for the classification of selected objects, the SVM classifier based on the SVM algorithm will be used [1 – 3]. Also, the HOG descriptor (Histogram of oriented gradients) realizing the counting of the number of gradient directions in local areas of the image will be used.

HOG descriptor allows to get a feature vector that describes the shape of objects in the image. This feature vector can be used for training in the development of various classifiers using the appropriate algorithms. The authors of the HOG descriptor proposed the use of a linear SVM classifier. A herewith the indicate the following disadvantages of this approach to the development of the SVM classifier [4]: the presence of the significant time expenses for both the formation of the histograms and the development of the SVM classifier based on the dataset received with calculating of these histograms.

The purpose of this work is the study of the dependency of the results of the SVM classifier development using the dataset formed on the basis of the HOG descriptor, from the number of the considered features of objects in this dataset. As part of this study, it is proposed to use the special tool known as the feature selection to determine the most informative features of objects of the training set, that should allow to reduce the time expenses on solving the objects detection problem while maintaining acceptable values of the classification quality indicators.

2. Decision of the objects detection problem

The decision of the object detection problem can be divided into the following steps.

1. The generation of features: the choice of those features that describe the object with sufficient completeness (within reasonable limits).
2. The feature selection: the selection of the most informative features for classification.
3. The classifier development: the choice of the decision rule according to which, based on the feature vector, the object is assigned to the particular class.
4. The assessment of the classification quality.

The descriptors solve the problem of the image description. The descriptor provides a description of the singular point, which determines the features of its neighborhood, and represents a numerical or binary vector of the certain parameters. The descriptor allows selecting the singular point from the set of all points in the image.

We used the HOG descriptor to describe the features of objects. HOG is a feature descriptor used to detect ethe objects in computer vision and image processing. The HOG descriptor technique counts occurrences of gradient orientation in localized portions of an image – detection window, or region of interest [4].

The main problem of application of the HOG descriptors is the huge number of the generated features that lead to the significant time expenses on the classifier development (in the context of the object detection problem). Therefore, it is necessary to make the feature selection.

We suggested to make the feature selection using the neighborhood component analysis (NSA) algorithm for classification [5 – 7]. This algorithm performs the feature selection for classification using the predictors (the inputs of the dataset) and the responses (the output of the dataset). It finds the feature weights by using a diagonal adaptation of the NCA with regularization. The weights of the irrelevant features will be close to zero.

This algorithm finds the neighborhood component analysis model for classification using Stochastic Gradient Descent (SGD). SGD is a simple yet very efficient approach to discriminative learning of linear classifiers under convex loss functions.

In the context of solving the problem of the objects detection, it will be necessary to define the classification rule, that is, the rule for assigning the object to the particular class. In the case of the binary classification, two classes are considered: "object is detected" and "object is not detected".

One of the algorithms that successfully solve this problem is the SVM algorithm used to develop the SVM classifier.

The main feature of the SVM classifier is the use of the kernel function which is applied to transfer the experimental dataset from the original feature space to a higher dimension space, in which the hyperplane separates the classes is built. Moreover, on both sides of the separating hyperplane, two parallel hyperplanes, defining the boundaries of the classes and located at the maximum possible distance from each other, are built.

As a result of the SVM classifier training, the separating hyperplane, which can be given by the following equation is defined [1 – 3]:

$$< w, z > + b = 0,$$

где w is the perpendicular vector to the separating hyperplane; b is the bias, z is the object.

A herewith, the support vectors that are closest to the hyperplane separating the classes and carry all the information about the classes separation are determined.

The classification decision that associates the object z with the class marked by "–1" or "+1" is made in accordance with the rule [1 – 3]:

$$F(z) = sign(< w, z > +b).$$

To assess the classification quality, usually the following indicators are used: overall accuracy, sensitivity, specificity, F-measure, the errors of the 1-st and 2-nd types. Also, it is useful to assess the number of the support vectors for SVM classifier.

3. Experimental studies

The experimental studies were performed with the INRIA Person Dataset. The characteristics of this dataset are shown in table 1.

To assess the quality of the obtained SVM classifiers, the hold-out validation method was used. A herewith, the proportion of the test set in the experimental dataset was equal to 0.2.

In these experiments, the parameters of the SVM classifier were taken by default equal to the following values: the value of the regularization parameter $C = 1$; the value $\sigma = 1$ for the radial basis kernel function.

The experiments were conducted using a PC with Windows 10 operating system, based on the Intel Core i3-4160 processor with 8 GB RAM. The MATLAB application package ver. R2017a was used as the computing runtime environment. During the experiments, we estimated the weights values of the features using the NCA algorithm. The results of calculations are presented graphically in figure 1.

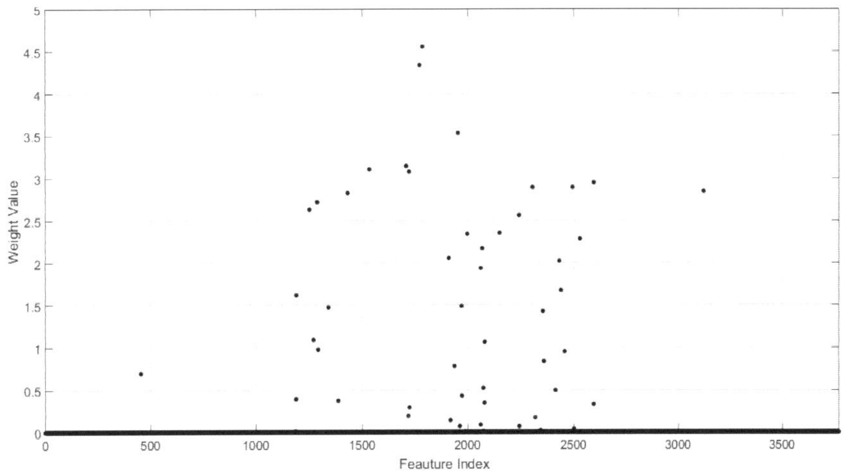

Figure 1. The feature weights

Figure 2 shows the dependence of the number of features on the weight threshold value, which was calculated in accordance with the formula for the terms of the geometric progression:

$$Q_n = 10^{-75} \cdot 100^{n+1}.$$

In accordance with the approach proposed by the authors of the HOG descriptor, a linear SVM classifier was trained. A herewith, only features that crossed the threshold Q_n participated in this training. Figure 3 shows the number of errors in the training and test sets with an increase in the weight threshold value, and as a result, with a decrease in the number of the considered features. The analysis of this figure allows to say that with a decrease in the number of the considered features, the large number of errors appear in the training set, which means the set becomes linearly inseparable. At the same time, with a small threshold value, that is, with a small filtering out of the features of objects, there is the acceptable generalizing ability of the trained model, consisting in a small number of errors in both the training and test sets.

Table 1. The characteristics of the experimental dataset

Feature	Value
The number of objects in the experimental data set	4172
The number of objects of class "+1" (that is, the number of targets)	2172
The number of objects of class " – 1" (that is, the number of objects of the background)	2000
The proportion of the test set in the experimental dataset	0.2
The number of objects in the training set	3338
The number of objects in the test set	834

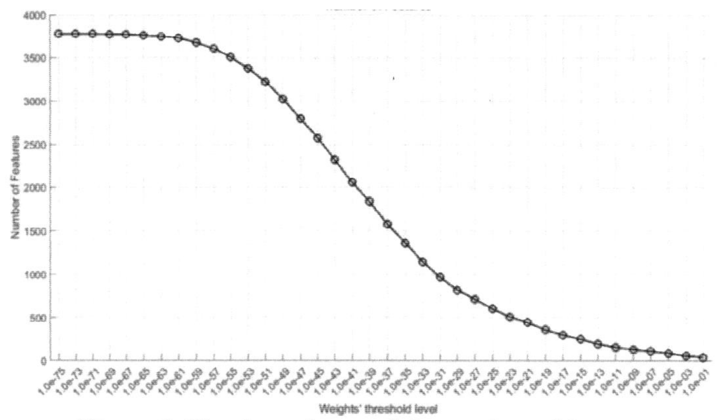

Figure 2. The dependency of the number of features from the threshold weight for the linear kernel function

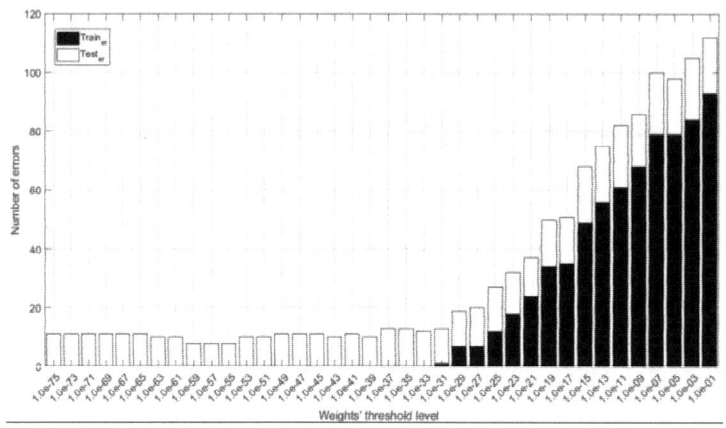

Figure 3. The dependency of the number of errors in the training and test sets from the threshold weight for the linear kernel function

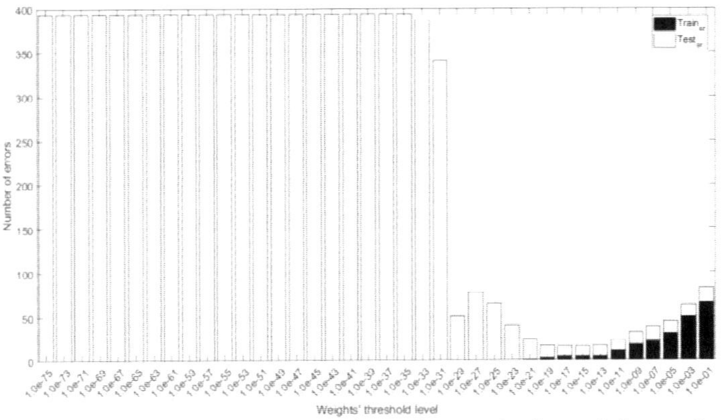

Figure 4. The dependency of the number of errors in the training and test sets from the threshold weight for the radial basis kernel function

Based on the assumption that the set becomes linearly inseparable, it was decided to train the SVM classifier with the nonlinear kernel function, namely, with the radial basis kernel function. The learning results are presented in the same format as for the linear kernel function in figure 4. With a small threshold, there is an obvious re-training of the model, which is characterized by a small number of errors in the training set and a large number of errors in the test set. Indeed, the corresponding number of support vectors (figure 5) for the analyzed range of weights indicates the unsatisfactory generalizing ability of the model and the obvious retraining.

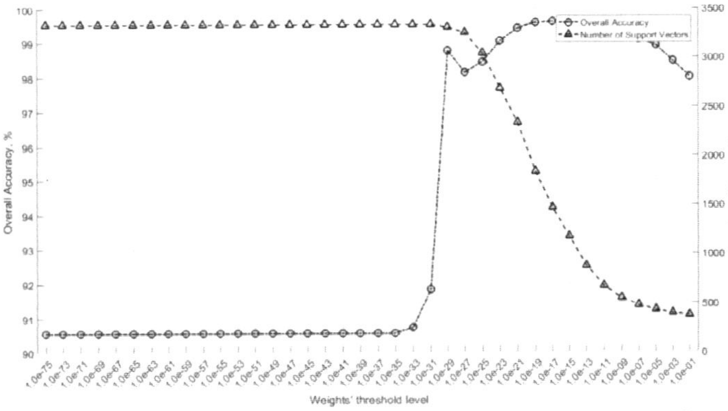

Figure 5. The dependencies of the overall accuracy and the number of support vectors from the threshold weight for the radial basis kernel function

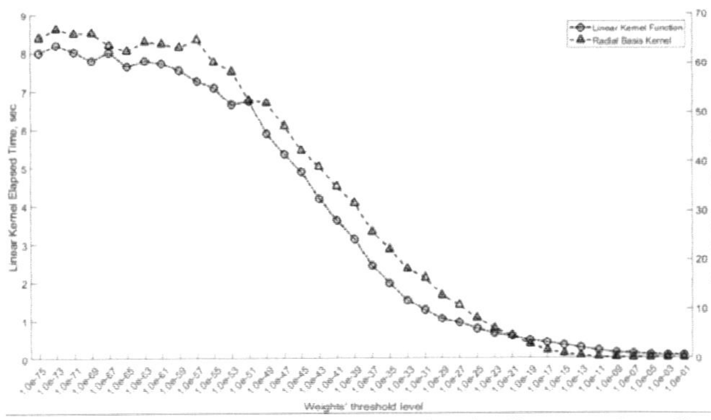

Figure 6. The dependency of the development time for the linear and nonlinear SVM classifiers from the number of the considered features of objects

On the basis of the obtained results, the following conclusion can be drawn: as the number of the considered features decreases in accordance with the weights obtained by the feature selection algorithm, the training set becomes linearly inseparable. The transition from the linear separability to the linear inseparability for the considered example is observed for the threshold value within the range from 1E–29 to 1E–17. The number of features decreases from 819 to 298. It should be noted that, for example, the results of the nonlinear SVM classifier at the threshold which equals to 1E–17, when the number of the considered features is 298, are not inferior to the results of the linear SVM classifier trained on the entire set of features.

Figure 6 shows the dependency graphs of the learning time for the linear and nonlinear SVM classifiers. The time of learning of the linear SVM classifier with the entire feature set was equal to 8 seconds. The time of learning of the nonlinear SVM classifier with the threshold weight value which equals to 1E–17 was equal to 1.7 seconds.

The studies conducted on the considered dataset prove the feasibility of reducing the number of features. The open libraries of machine learning and computer vision use much larger datasets for learning. When forming these datasets, some initial set of images containing objects of classes is subject to various distortions. The purpose of such distortions is an attempt to provide for various options for the location and shape of some object in the image. In this case, the distortion affects the number of informative features. It should be noted that the main goal of the open libraries developers is to develop the universal tools for solving detection problems that meet the most frequent requirements of users. However, if the goal is to develop a proprietary software solution for working in the specific conditions and with the common format for representation of the objects classes, it is necessary to consider the possibility of accelerating both the calculation of the histogram of the oriented gradients and the development of the SVM classifier, which was shown in this experimental study.

4. Conclusion

Experimental studies have confirmed the high efficiency of application of the special tool known as the feature selection using the neighborhood component analysis algorithm for classification to determine the most informative features of objects. It is shown that this tool allow to reduce the time expenses on solving the objects detection problem while maintaining acceptable values of the classification quality indicators.

5. References

[1] VapnikV 1998 *Statistical Learning Theory*

[2] Demidova L, Sokolova Yu and Nikulchev E 2015 Use of fuzzy clustering algorithms ensemble for SVM classifier development *International Review on Modelling and Simulations* **8(4)** 446-457

[3] Demidova L , Nikulchev E and Sokolova Y 2016 The SVM Classifier Based on the Modified Particle Swarm Optimization *International Journal of Advanced Computer Science and Applications* **7**(2) 16-24

[4] Navneet Dalal and Bill Triggs 2005 *Proceedings of the IEEE Conference on Computer Vision and Pattern Recognition* **II**

[5] Goldberger J, Roweis S T, Hinton G E, Salakhutdinov R 2004. *NIPS'04*

[6] Yang Z, Laaksonen J 2007 Regularized 2007 *Image Analysis: Lecture Notes in Computer Science* **4522**

[7] Yang W, Wang K, Zuo W 2012 *Journal of Computers* **7**

IOP Conf. Series: Materials Science and Engineering **476** (2019) 012008 doi:10.1088/1757-899X/476/1/012008

An Intellectual Approach to Segmentation of the Satellite Images

L A Demidova[1,2] **and R V Tishkin**[3]

[1]Ryazan State Radio Engineering University, Ryazan, Russia
[2]Moscow Technological Institute, Moscow, Russia
[3]Limited Liability Company "Areal-98", Moscow, Russia

E-mail: demidova.liliya@gmail.com

Abstract. The applicability of the clustering algorithms under uncertainty conditions for the image segmentation from the remote sensing satellites has been shown. Approaches to expanding the characteristics' space to improve the quality of the image segmentation have been considered. An intellectual approach to improving the quality of the image segmentation, which consists in combining the results of the image segmentation from the different clustering algorithms under conditions of uncertainty through their use in the development of the SVM classifiers and their ensembles, has been formulated. The results of solving the problem of the image segmentation based on the clustering algorithms under uncertainty conditions have been presented.

1. Introduction

Image segmentation plays an important role in Earth remote sensing systems and is used to implement the scene recognition and object selection. Typically, a digital image is represented as a rectangular grid of pixels, which causes considerable inconvenience in the high-level analysis. The purpose of segmentation is to simplify the description of the image by moving from a representation in the form of a set of the raster points to a representation in the form of a set of the objects. This transition is based on the selection of areas with the similar visual characteristics.

Let for the original image we have l characteristics (brightness, color, texture, dispersion, etc.), which allow to localize the objects of interest. The segmentation problem can be reduced to the clustering problem of the objects' set in the l-dimensional space, if it is possible to express for each pixel the numerical values of the characteristics on the scale of intervals or relations. It is worth noting that the uncertainty due to the inability to unambiguously classify the objects depicted is inherent in the segmentation problem. So, it is advisable to apply the clustering algorithms under uncertainty conditions and implement a particular interpretation of uncertainty. In this case, the same image pixel can belong to the several clusters simultaneously, which provides the flexibility for further image analysis.

The most well-known clustering algorithms under uncertainty conditions are the fuzzy c-means algorithm (FCM algorithm) and its modifications: the possibilistic c-means algorithm (PCM algorithm) and the possibilistic-fuzzy c-means algorithm (PFCM algorithm) [1 – 3]. A herewith, these algorithms can operate with type-1 fuzzy sets and type-2 ones, which work with the footprint of uncertainty and allow describing the uncertainty more correctly [4].

Under the clustering problem scientists understand the finding of a partition of the original set of objects that form the structure of clusters present in the analyzed data. Depending on the clustering algorithm used, this problem is reduced to finding the crisp membership of objects (for example, pixels) to clusters (in the case of the classical crisp clustering algorithms, for example, the k means algorithm) or degrees of the ownership (typicality) of objects to clusters, which together define a fuzzy (possibilistic, possibilistic-fuzzy) partition of the original set of pixels.

In general, the clustering problem has the form: to determine such partition $P(R)=\{R_k \mid R_k \subseteq R\}$ of the pixels' set of image R for a given number c of clusters R_k ($k=\overline{1,c}$), which provides an extremum of a certain objective function $f(P(R))$ among all partitions [1]. For example, a fuzzy partition of the image pixels is a system of fuzzy subsets $P(R)=\{R_k \mid R_k \subseteq R\}$, if the condition $\bigcup_k R_k = R$ ($R_k \in R$) is fulfilled.

To improve the quality of the image segmentation, it is reasonably, firstly, to apply the different clustering algorithms to the image, and, then, to apply an approach, realizing modern intellectual technique that allows to take into account all clustering results simultaneously by the best way.

2. FCM algorithm

For calculating the distance $d(x_i,x_q)$ between the objects x_i and x_q in the l-dimensional characteristics' space in the clustering algorithms under uncertainty conditions the following function, based on Euclidean metric, can be used:

$$d(x_i,x_q)=(\sum_{j=1}^{l}(x_i^j-x_q^j)^2)^{0.5} \quad (1)$$

The FCM algorithm (fuzzy c-means) implements the fuzzy interpretation of uncertainty and belongs to iterative algorithms that compute the values of the membership functions to the clusters and the centers' coordinates in accordance with the values of the membership functions [1, 2].

The FCM algorithm performs the minimization of the objective function [1, 2]:

$$J(U,V)=\sum_{r=1}^{c}\sum_{i=1}^{n}(u_r(x_i))^m \cdot d^2(v_r,x_i) \quad (2)$$

given that

$$\sum_{r=1}^{c}u_r(x_i)=1 \ (c\in N \text{ and } c>1;\ i=\overline{1,n}), \quad (3)$$

where $U=[u_r(x_i)]$ is a fuzzy c-partition of the objects' set X on the base of the membership functions $u_r(x_i)$, determining the degree of membership of the i-th object to the r-th cluster; $V=(v_1,...,v_c)$ is the clusters' centers' vector; $d(v_r,x_i)$ is the distance between the cluster center v_r and the object x_i in accordance with (1); m is a fuzzyfier ($m\in R$, $m>1$); c is the clusters' quantity; n is the objects' quantity; $i=\overline{1,n}$; $r=\overline{1,c}$.

The FCM algorithm involves the following steps [2].

1. Initialization of c-partition $U=[u_r(x_i)]$, which satisfies (3).
2. Calculation of the centers' coordinates:

$$v_r^j=(\sum_{i=1}^{n}u_r(x_i)^m \cdot x_i^j)\Big/\sum_{i=1}^{n}u_r(x_i)^m \quad (4)$$

3. Calculation of the membership functions values:

$$u_r(x_i)=\frac{1}{\sum_{t=1}^{c}\left(d(v_r,x_i)/d(v_t,x_i)\right)^{2/(m-1)}} \quad (5)$$

4. Steps 2 and 3 are repeated until a predetermined quantity of iterations H is reached or a predetermined accuracy is reached: $|J(U,V)-J'(U,V)|\le\varepsilon$, where $J(U,V)$, $J'(U,V)$ are the values of the objective function (2) in two successive iterations.

The membership functions $u_r(x_i)$ play the role of the weight coefficients, determining the degree of a pixel contribution to the cluster center assessment and, accordingly, the degree of belonging of the i-th pixel to the r-th cluster. The degree of a pixel contribution depends on the choice of the fuzzifier, which controls the degree of the "fuzziness" of the segmentation.

For small values m, a fuzzy partition degenerates into a crisp one, if $m\to\infty$, then the membership functions of pixels to each cluster become equal to $1/c$, that is, each pixel belongs to each cluster with equal probability. Usually, a value of 2 is selected as the value of the fuzzifier.

When using the FCM algorithm, the local-optimal fuzzy partition is calculated, which is described by a set of the membership functions, and the coordinates of the clusters' centers. To obtain the adequate results of the fuzzy clustering, it is necessary to repeatedly perform the FCM algorithm for a given number of clusters for the different initial fuzzy partitions in order to make a final decision on the desired fuzzy clustering. Also, it is possible to use any evolutionary optimization algorithm, for example, the genetic algorithm or the particle swarm optimization algorithm.

As an indicator of the partition quality it is recommended to use the index XB:

$$XB=\frac{J(U,V)}{n\cdot \min\limits_{k\neq t} d(v_k,v_t)}, \tag{6}$$

where $J(R,U)$ is an objective function (2).

The index XB takes into account both the fuzzy degrees of the objects belonging to the centers of the clusters and the geometrical arrangement of the centers of the clusters and objects, which in the most cases allows obtaining the adequate segmentation results. As a result, the centers of the desired clusters "tightened" to the aggregation of objects, while ensuring minimal similarity between the characteristics of pixels belonging to different clusters.

The FCM algorithm does not always accurately estimates the centers' coordinates due to the use of possible enforcement, determined in (3). The membership functions calculated based on the constraint (3) determine the separation degrees rather than the typicality degrees for pixels. Therefore, atypical pixels (pixel-noise) located at the same distance from the real cluster centers can strongly influence the estimates of the coordinates of the cluster centers, and, consequently, the final result of clustering. One of the approaches to reduce the noise effect (the influence of the atypical pixels on clustering results) is based on the use of the weighting factors that are inversely proportional to the distance of a pixel from the center of the cluster.

The PCM algorithm can be used to reduce the effect of the atypical objects on the clustering results [2, 3]. The PCM implements the possibilistic interpretation of uncertainty and belongs to iterative algorithms that compute the typicality functions to the clusters and the centers' coordinates in accordance with the values of the typicality functions [2, 3].

The use of the PCM algorithm is effective if the set of the image pixels contains pixel noise. However, the use of the PCM algorithm can lead to the formation of the coincident clusters, since the functions of the typicality of pixels to a certain cluster do not depend on the distances of the pixels to other clusters. In this case, a reasonable approach will be the simultaneous consideration of both cluster relativity and cluster typicality of pixels, that is realized in the PFCM algorithm [2].

These algorithms have the different objective functions and implement the uncertainty accounting differently. Therefore, they can give different the clustering results for some objects (pixels).

In this regard, it is advisable to implement an approach that implements the ensemble of the results from different clustering algorithms.

3. Expansion of the characteristics' space

The use of the FCM algorithm in the classic version described above, that is, directly to the original set of the image pixels to perform their segmentation, is accompanied by ignoring the information about the location of pixels on the original image, that can lead to the formation of many small, significantly distant from each other areas related to one cluster. A herewith, the neighboring pixels can be assigned to different clusters, due to the possible noise in the image.

In the experiments we used the set of images obtained from the Resurs-DK spacecraft and analyzed the results of the segmentation of these original images based on the FCM algorithm in the classic version. In many experiments we found that the original images were fragmented into several disconnected areas, that can make it difficult to further high-level processing.

To eliminate this drawback (the fragmentation of images into unrelated areas) the so-called expansion of the characteristics' space can be implemented [5]. In the classic version of the fuzzy clustering algorithm (that is, without expanding the characteristics' space), each pixel is associated with a vector r_i composed of the characteristic values of only this pixel. When performing the expansion of the characteristics' space, the characteristic values of the pixels in a four- or eight-connected domain adjacent to the considered one are added to the vector r_i (Fig. 1).

a) b)

Figure 1. The domains: a) the four-connected domain; b) the eight-connected domain

A comparative analysis of the clustering results with expansion of the characteristics' space did not reveal the significant differences in the segmentation quality when the four- and eight-connected domains were used. A herewith, the processing speed of a four-connected domain is substantially (up to one and a half) higher than the processing speed of an eight-connected one. In this regard, the expansion of the characteristics' space on the four-connected region is preferable to use for the image segmentation.

The segmentation results of the original images based on the FCM algorithm with expansion of the feature space using a four-connected domain show, that the expansion of the characteristics' space made allows to obtain the unfragmented selection of the observed objects. Thus, the expansion of the characteristics space provides more adequate segmentation results.

4. Experimental studies

The experimental studies were performed for the identification problem of the cloud formations in the satellite images. In this subject area, the clustering should be carried out in the HSV color space (Hue, Saturation, Value).

Since a person can easily recognize the cloudiness in the satellite images, the results of manual separation of the cloud formations can be taken as the segmentation standard. The original satellite image and the results of the manual processing are presented in Fig. 2 and 3, respectively.

To visualize the results of the work of the fuzzy clustering algorithm described above, it is necessary to convert the resulting fuzzy partition into a crisp one.

A herewith, the following defuzzification rule can be used:

"If $u_k(x_i) > u_t(x_i)$ for $t = \overline{1,c}$, $k = \overline{1,c}$ и $k \neq t$, then pixel x_i strictly belongs to the cluster k ".

The segmentation results of the original image (Fig. 2) based on the FCM algorithm are shown in Fig. 4.

Figure 2. The original image

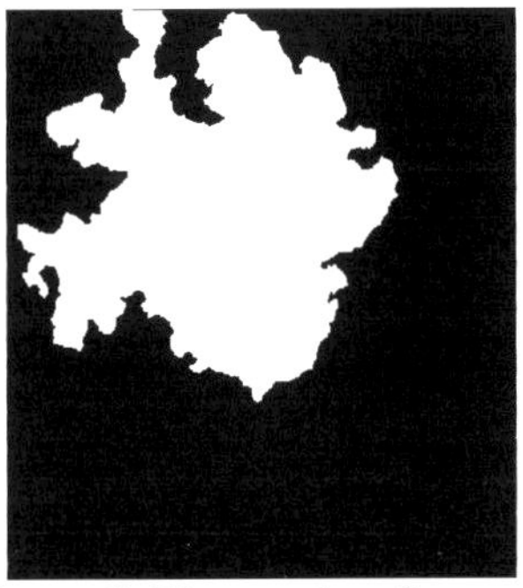

Figure 3. The results of manual selection of the cloud formations

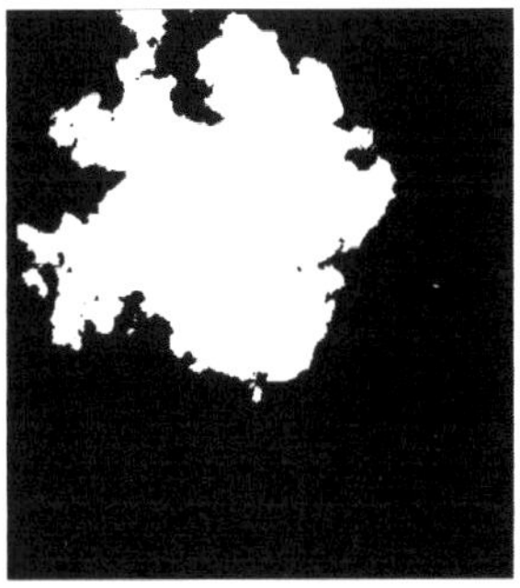

Figure 4. The segmentation results of the cloud formations based on the FCM algorithm

To evaluate the accuracy characteristics of the FCM algorithm, when performing the image segmentation 50 experiments were conducted. Based on the pixel-by-pixel comparison of the segmentation results based on the FCM algorithm with the results of manual selection of the cloud formations, the averaged percentages of the true and incorrectly classified pixels to the total number of pixels in the original image, were calculated. A herewith, 94,73 % of pixels were correctly classified, 1,52% of pixels were mistakenly referred to the cloud cluster, and 3,75 % of pixels were mistakenly not referred to the cloud cluster. The obtained assessments of the accuracy characteristics indicate the feasibility of applying clustering algorithm to the image segmentation problems. For the original image of 600x700 pixels in size, the segmentation time based on the FCM algorithm averaged over 50 experiments was 10.54 seconds.

Due to the independence of the calculations for the image pixels in the FCM algorithm, the processing speed can be significantly with the use of parallel computing technology.

The same close segmentation results can be obtained with other clustering algorithms under uncertainty conditions. However, for some pixels, the segmentation results may be different due to the specificity of the applied clustering algorithms under uncertainty conditions.

To improve the quality of the image segmentation and solve the problem of the ambiguous classification of some pixels by different clustering algorithms under uncertainty conditions, it is possible to use voting technique according to the majority vote. Also, it is possible to use more modern intellectual approach to making-decision based on application of the SVM classifier (Support Vector Machine) [2, 6-8], which allows to realize the binary pixel classification [7]. In this case, the well-classified pixels (i.e., the pixels equally classified by the different clustering algorithms) should be used to form the experimental dataset which is applied for creation of the training and test sets for development of the SVM classifier. The developed SVM classifier can be used to classify the poorly classified pixels (i.e., pixels differently classified by the different clustering algorithms under uncertainty conditions). The ways of forming of the experimental dataset were described in [2].

Moreover, the use of the ensembles of the SVM classifiers is possible [3]. Application of the SVM classifiers and their ensembles allow improving the segmentation quality by 2-5%.

5. Conclusion

The experimental studies show the high efficiency of the clustering algorithms under uncertainty conditions in processing images from remote sensing spacecraft. A herewith, the expansion of the characteristics' space of to the four-connected domain allows to significantly improve the results of the image segmentation at an acceptable time cost. The use of the SVM classifiers and their ensembles is appropriate in solving the problem of the ambiguous classification of a portion of image pixels by the different clustering algorithms under uncertainty conditions. Also, the random forest algorithm (RF algorithm) or the artificial neural networkscan be applied in solving the problem of the ambiguous classification, since they use the general paradigm of the machine learning algorithms. Therefore, the purpose of the further research is to implement a new tool for the data classification and a comparative analysis of the classification results in the case of the ambiguous classification fulfilled by the different clustering algorithms under uncertainty conditions.

6. References

[1] Nascimenko S, Mirkin B and Moura-Pires F 2000 *Fuzzy Systems* **1**.
[2] Demidova L, Sokolova Yu and Nikulchev E 2015 Use of fuzzy clustering algorithms ensemble for SVM classifier development *International Review on Modelling and Simulations* **8(4)** 446-457
[3] Krishnapuram R and Keller J 1993 *IEEE Transactions on Fuzzy Systems* **1**.
[4] Rhee F C-H 2007 *IEEE Computational intellegence magazine* **2(1)**.
[5] Krishna Priya R, Thangaraj Dr C and Kesavadas Dr C 2011 *IJCSI International Journal of Computer Science Issues* **1(1).**
[6] Demidova L , Nikulchev E and Sokolova Y 2016 The SVM Classifier Based on the Modified Particle Swarm Optimization *International Journal of Advanced Computer Science and Applications* **7**(2) 16-24
[7] Saha I, Maulik U, Bandyopadhyay S and Plewczynski D 2012 *IEEE Geoscience and remote sensing letters* **9**.
[8] VapnikV 1998 *Statistical Learning Theory.*

IOP Conf. Series: Materials Science and Engineering **476** (2019) 012009 doi:10.1088/1757-899X/476/1/012009

Improving the Efficiency of Aviation Products Design Based on International Standards and Robust Approaches

A Dmitriev[1, 2] and T Mitroshkina[2]

[1]Samara National Research University named after academician S.P. Korolev (Samara University), Samara 443086, Russian Federation
[2]Samara Region State Academy (Nayanova), Samara 443010, Russian Federation

E-mail: dmitriev57@rambler.ru

Abstract. The article considers the possibility of using QFD, robust design and other modern approaches for designing innovative aviation products. As an example, the projected unmanned aerial vehicle and engine, as its key part, are considered. Based on expert evaluation and analysis of benchmarking data, the main consumer requirements for UAV and a small-size gas turbine engine were identified. It is also necessary to pay attention to such general requirements as low noise, low emission of harmful substances (CO and NOx). The use of international standards of the ISO 16355 series as a methodology and basis for improving design efficiency is discussed. The results and conclusions can be applied to various types of aviation products, as well as used in other industries.

1. Introduction

Aviation industry is one of the strategic in Russia. The goal of the state program of the Russian Federation "Aviation Industry Development, 2013-2025" is "to create a highly competitive aviation industry and consolidate its position in the world market as the third manufacturer in terms of production of aviation equipment." High efficiency of the aviation products development based on meeting the requirements of standards and implementation of quality management methods is the most important direction in improving the competitiveness of products and Russian enterprises of the aviation industry.

Implementing the requirements of international standards, as well as applying quality management methods, such as “quality function deployment” (QFD), “failure mode and effect analysis” (FMEA), robust design and others are organizational innovations that aimed is helping organizations develop or improve the basic management system to ensure continuous improvement of the organization [1].

The above fully applies to the design and manufacture of unmanned aerial vehicles. This type of aircraft is used for video monitoring of the terrain in real time, for the tasks of oil and gas production enterprises, pipeline transportation, in power engineering and communications, medicine, for the tasks of the Ministry of Emergency Situations and so on. Due to the high extent and territorial extent of the objects of observation, aerial monitoring is the most effective means of monitoring and remotely collecting data on their condition.

The relevance of the topic of work is that the high efficiency of the development of aviation products and quality management based on robust design methods is the most important direction in improving the competitiveness of products. Otherwise as a result of a superficial approach to customer requirements, it may ultimately lead to unnecessary costs, and as a result, a decrease in the competitiveness of the manufactured product.

2. Quality function deployment

Quality function deployment (QFD) is a method to assure customer or stakeholder satisfaction and value with new and existing products by designing in, from different levels and different perspectives, the requirements that are most important to the customer or stakeholder. These requirements should be well understood through the use of quantitative and non-quantitative tools and methods to improve confidence of the design and development phases that they are working on the right things. In addition to satisfaction with the product, QFD improves the process by which new products are developed [2].

Reported results of using QFD include improved customer satisfaction with products at time of launch, improved cross-functional communication, systematic and traceable design decisions, efficient use of resources, reduced rework, reduced time-to-market, lower life cycle cost, improved reputation of the organization among its customers or stakeholders.

Since its inception in 1966, QFD has broadened and deepened its methods and tools to respond to the changing business conditions of QFD users, their management, their customers, and their products. Those who have used older QFD models will find these improvements make QFD easier and faster to use. The methods and tools shown and referenced in the standard represent decades of improvements to QFD; the list is neither exhaustive nor exclusive. Users should consider the applicable methods and tools as suggestions, not requirements.

Current Japanese users of QFD concepts include Nissan, Toyota, Komatsu, Nippondenso and Honda. In the United States of America users include Ford, GM, Chrysler, DEC, TI, 3M, HP, AT&T Bell Labs, NovAtel, Xerox, Exxon and Dow.

2.1. Standardisation: Transitioning from JIS Q 9025 to ISO 16355

Since the first published article in 1966 [3], QFD has grown from the strong roots of Japanese quality management methods and tools into the Japanese standard JIS Q 9025:2003 [4], which is based on the Comprehensive QFD model developed by Akao and others [5] W. Edwards Deming said that knowledge comes from "outside" the system [6], and as QFD has extended its influence globally over the past thirty years, so has QFD absorbed best practices from worldwide practitioners. Based on papers and case studies, the ISO 16355 series of standards brings these Japanese and global methods and tools together to update the new product development process so that it better aligns with today's and tomorrow's customers, suppliers, and competitors [7].

These global methods and tools not only strengthen the classical Comprehensive QFD model, they also integrate with modern strategic planning, market and consumer research, voice of customer analysis to uncover unspoken needs, and innovation. Then, after product launch, the ISO 16355 provides guidance for ongoing support and environmental sustainability.

The ISO 16355 includes 8 parts: 1. General principles of QFD; 2. Acquisition of voice of customer/stakeholder (VOC/VOS) – non-quantitative approaches; 3. Acquisition of VOC/VOS – quantitative approaches; 4. Analysis of VOC/VOS; 5. Strategy and Translation of VOC/VOS into engineering solutions and cost planning; 6. Optimization – robust parameter design; 7. Optimization – tolerance design; 8. Commercialization and life cycle. Therefore the greatest attention is given to acquisition and analysis of voice of customer/stakeholder (VOC/VOS).

First part of ISO 16355 "General principles of QFD" describes the QFD process, its purpose, users, and tools. It contains overview and flow of modern QFD and robust design (Comprehensive QFD for quality, technology, cost, and reliability; 4-Phase QFD for auto parts suppliers; Blitz QFD® to fast-track critical customer needs), 25 figures and tables and 125 published references. It is not a management system standard.

So it does not provide requirements or guidelines for organizations to develop and systematically manage their policies, processes, and procedures in order to achieve specific objectives [2]. The contents of first part of ISO 16355 are shown in Figure 1.

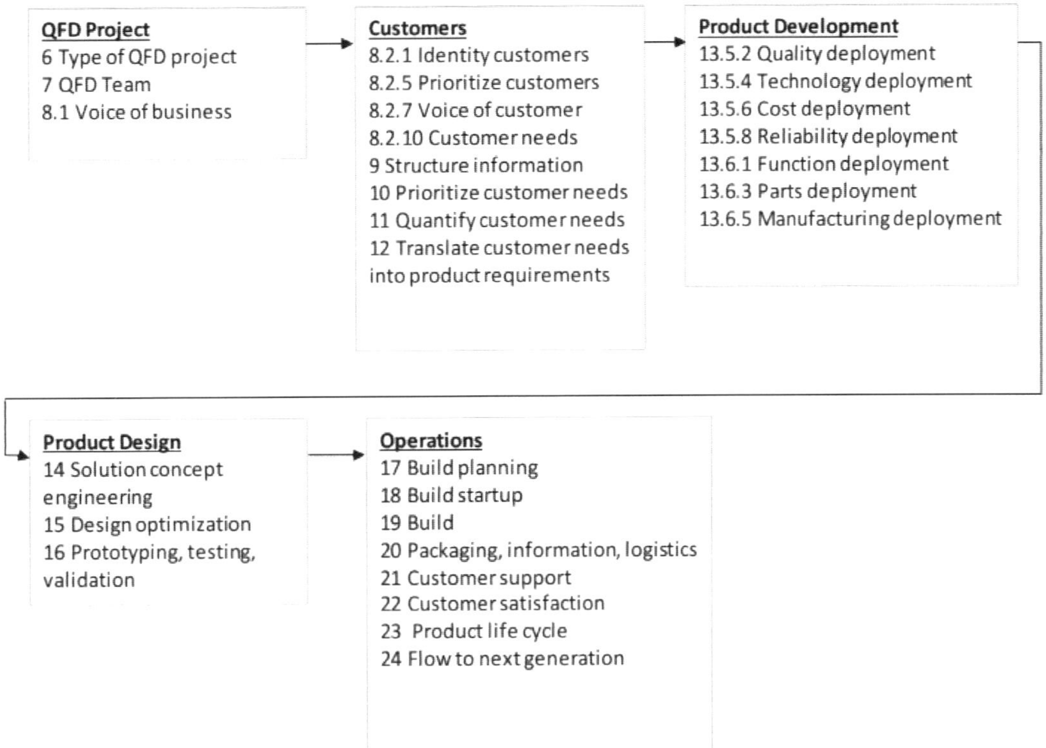

Figure 1. Contents of ISO 16355-1.

Users of this part of ISO 16355 will include all organization functions necessary to assure customer satisfaction, including business planning, marketing, research and development (R&D), engineering, information technology (IT), manufacturing, procurement, quality, production, sales, service, packaging and logistics, support, testing, regulatory, and other phases in organizations.

3. Application of QFD and robust approaches for unmanned aerial vehicles design

International Civil Aviation Organization (ICAO) recognizes many categories of aircraft, among them balloons, gliders, aero planes and rotorcrafts. Aircraft can be land, sea or amphibious. Whether the aircraft is manned or unmanned does not affect its status as an aircraft. Each category of aircraft will potentially have unmanned versions in the future [8]. This point is central to all further issues pertaining to UA and provides the basis for addressing airworthiness, personnel licensing, separation standards, etc. Unmanned aircraft systems (UAS) are a new component of the aviation system, one which ICAO, States and the aerospace industry are working to understand, define and ultimately integrate. These systems are based on cutting-edge developments in aerospace technologies, offering advancements which may open new and improved civil/commercial applications as well as improvements to the safety and efficiency of all civil aviation. The safe integration of UAS into non-segregated airspace will be a long-term activity with many stakeholders adding their expertise on such diverse topics as licensing and medical qualification of UAS crew, technologies for detect and avoid systems, frequency spectrum (including its protection from unintentional or unlawful interference), separation standards from other aircraft, and development of a robust regulatory framework.

3.1. Study of robust methods for designing and improving UAVs

Model of 4-Phase QFD for auto parts suppliers ISO 16355 [2] and previous work experience [1, 9, 10] was used to design a UAV, the purpose of which is to solve civil tasks: monitoring problem areas of oil companies' main pipelines, inspecting hydraulic structures, identifying unauthorized dumps, monitoring the extraction of natural resources, and more.

The basic requirements of the consumer are: payload > 5 kg, working height from 1000 to 5000 meters, communication distance> 100 km, stable flight capability> 15 hours and flight range > 2000 km. Conducted collection and analysis of consumer requirements with the participation of experts from the Samara National Research University. The method of pairwise comparisons was used to collect and evaluate opinions. The results of assessing the relative importance of the requirements are presented in Table 1 and Figure 2.

Table 1. UAV customer requirements table.

#	UAV customer requirements	Weight, %
1	Maximum communication range (over 100 km)	13,0
2	The ability to control a long distance	12,5
3	Flight range (more than 1920 km)	11,8
4	The possibility of stable flight (more than 15 hours)	11,3
5	Maximum area of observation of the device with good image quality	10,9
6	Reusability	9,1
7	Maximum payload (more than 5kg)	8,9
8	Working height (from 1000 to 5000 m)	7,4
9	The minimum mass of the UAV (30 kg)	7,1
10	The ability to quickly move	3,4
11	Low cost	2,8
12	Fast assembly / disassembly and easy transportation	1,7

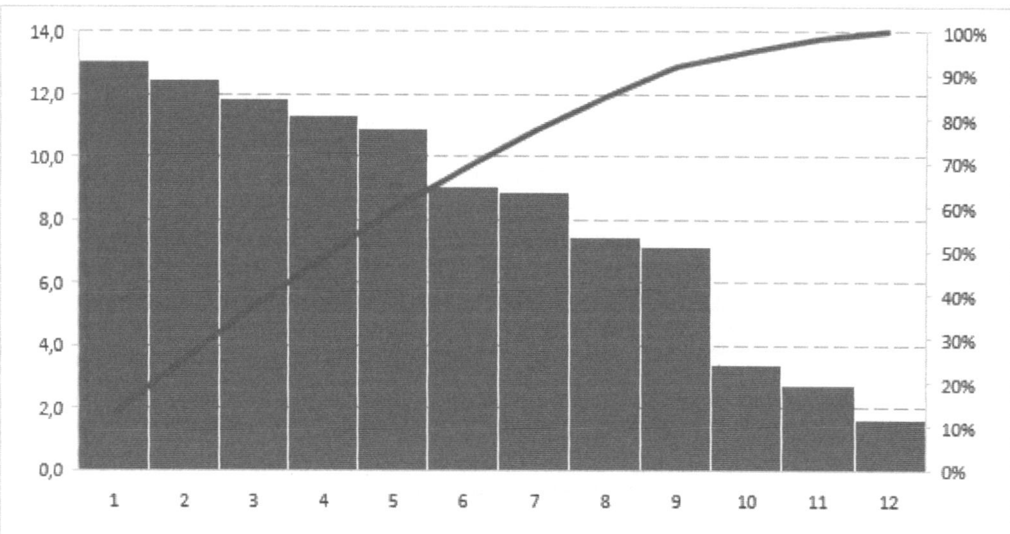

Figure 2. Pareto diagram for UAV customer requirements weight.

Thus, the most significant indicators are the maximum communication range, the ability to control at a long distance and the flight range. It is these parameters determine the radius of action of the aircraft, the area available to its review directly depends on them, which is very important both for military intelligence and for scientific observations. The most important requirements taking into account the results of benchmarking are the maximum communication range and the ability to control at a long distance.

At the first phase of QFD, the requirements of consumers using the house of quality (HoQ) were transferred to the flight performance of the UAV: Maximum take-off weight, Mass of payload, Relative payload ratio, Cruising speed, Maximum flight speed, Radius of action, Video transmission distance, Flight duration hour, Practical ceiling, Limit wind speed at the surface of the Earth at the start, Guaranteed resource, Viewing angle, Camera rotation angle, Deployment time.

Table 2. UAV Flight performance table.

#	UAV Flight performance	Weight, %
1	Radius of action	14,05
2	Maximum take-off weight	12,63
3	Video transmission distance	12,36
4	Mass of payload	9,55
5	Cruising speed	8,33
6	Maximum flight speed	8,33
7	Practical ceiling	7,95
8	Viewing angle	5,81
9	Camera rotation	5,81
10	Flight duration	5,56
11	Guaranteed resource	5,13
12	Deployment time	2,41
13	Relative payload ratio	1,10
14	Limit wind speed at the surface of the Earth at the start	0,98

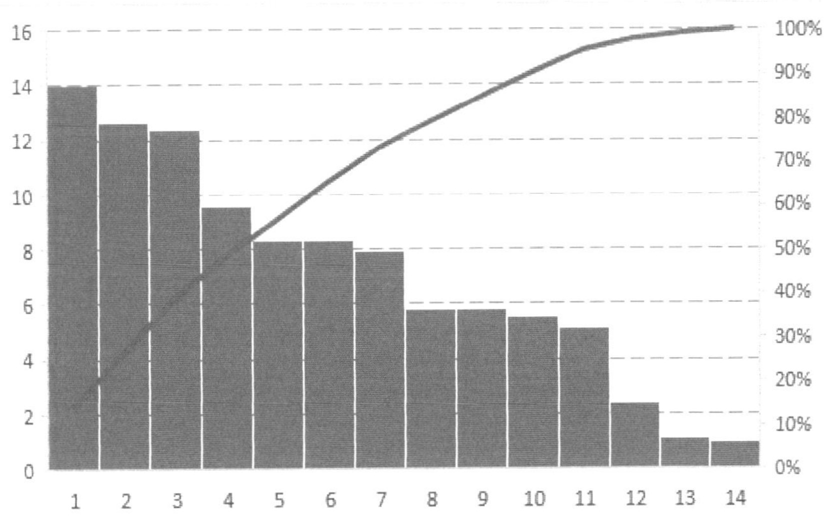

Figure 3. Pareto diagram for UAV Flight performance weight.

The flight performance characteristics that are most important for the implementation of customer requirements are Radius of action, Maximum take-off weight, Video transmission distance.

Then the technical characteristics of the UAV units are considered: Glider, Transceiver with rheostat-contactor control system, Target load, Power load, Energy sources. As a result of building a

House of Quality and FMEA, the most important characteristics of the components of an unmanned aerial vehicle have been identified, which need to be improved in the first place: telemetry and autopilot transceiver (weight 20.52%); opto-electronic system (weight 15.70%); relative thickness of the airframe shell (weight 12.98%).

Using methods Taguchi and experiments conducted in the universal software system ANSYS product Discovery Live, carried out parametric design of UAVs. As a result of the robust design, the optimal configuration of control factors (UAV take-off mass; wing span; wing area) was found, at which the signal-to-noise ratio takes the maximum value.

3.2. QFD to improve and study the small-size gas turbine UAV's engine

Additionally, the possibility of implementing QFD to improve and study the small-sized gas turbine engine of an unmanned aerial vehicle was considered.

A gas turbine engine is a type of heat engine that operates according to a certain principle. The principle of operation of the engine is to convert gas energy into mechanical work. The gas is compressed and heated before starting. A distinctive feature of a gas turbine engine from a piston engine is that all processes occur in a stream of moving gas.

As an object of study of this work, the TJ100 small-sized turbojet engine, which is a sub-type of small-sized gas turbine engines, was chosen. This small jet engine was developed for light aircraft, for example for light and ultra light sport aircraft, gliders equipped with an auxiliary engine, manned vehicles of the experimental category and various UAVs.

When developing unmanned aerial vehicles, it is necessary to ensure that the product conforms to all the conditions of the customer's technical specification. First of all, this applies to aircraft engines, the requirements for which can be divided: into general and to a specific type of engine installed on a specific UAV. General requirements for small-size gas turbine engine UAV include [10]:

1. Safety requirements: engine safety; stable operation of the power plant; stable performance.
2. Destination requirements: ensuring thrust on take-off mode; ensuring thrust in cruising flight conditions; ensuring specific fuel consumption in cruising flight conditions.
3. Design requirements: compliance with restrictions on size; compliance with weight limits; modularity.
4. Technical requirements: high reliability; high durability.
5. Operational requirements: high maintainability; ease of maintenance and repair.
6. Production and technological requirements: ease of manufacture and assembly; use of available materials.
7. Special Requirements: low noise; low emission of harmful substances (CO and NOx).
8. Technical and economic requirements: low cost of manufacturing a serial sample; low operating and maintenance costs.

Based on expert evaluation and analysis of benchmarking data, the main consumer requirements for a small-size gas turbine engine were identified: stable operation of the power plant, provision of thrust in take-off mode, provision of thrust in cruising flight conditions, provision of specific fuel consumption in cruising flight conditions, use of available materials, low cost of manufacturing. It is also necessary to pay attention to such general requirements as low noise, low emission of harmful substances (CO and NOx).

According to the results of QFD for the design of a small-size gas turbine engine, priority technical characteristics were determined (engine weight; total mass of the power plant and fuel consumed during the flight; engine thrust in take-off mode; hourly fuel consumption in the engine), as well as the most important indicators of the efficiency of engine components (full engine recovery ratio in the input device; full engine recovery ratio in the combustion chamber; nozzle speed ratio; ent complete combustion of the fuel in the combustion chamber).

4. Conclusion

The paper presents examples of the implementation of QFD in the development of UAVs and small-sized gas turbine engines of UAVs. The use of QFD in the development of innovative aviation products requires the involvement of experts and knowledge of statistical methods and robust design approach. An additional advantage is the possibility of conducting experiments using mathematical and virtual models in information systems, for example, ANSYS Discovery Live. The ISO 16355 series of standards are not standards for management systems and do not contain ready-made procedures for implementation. At the same time, the application of the tools given in the ISO 16355 series of standards allows the company to increase the efficiency of design and development of innovative products and processes.

References

[1] Dmitriev A Y 2016 *Robastnoye proyektirovaniye i tekhnologicheskaya podgotovka proizvodstva izdeliy aviatsionnoy tekhniki* (Samara: Samara University) [In Rus]
[2] ISO 16355-1:2015 2015 *Applications of statistical and related methods to new technology and product development process - Part 1: General principles and perspectives of Quality Function Deployment (QFD)* (International Standards Organization, Geneva, Switzerland)
[3] Glenn Mazur 2017 Transitioning from JIS Q 9025 to ISO 16355 *International Symposium on QFD (ISQFD)*
[4] Oshiumi K 1966 Perfecting Quality Assurance System in Plants *Quality Control*, pp 62-67.
[5] Japanese Industrial Standard 2003 JIS Q 9023:2003(E) Performance improvement of management systems – Guidelines for managemant by policy (Japanese Standards Association)
[6] Akao Y. and Mazur G H 2003 The leading edge in QFD: past, present and future *International Journal of Quality & Reliability Management* **20**(1) 20-35
[7] Deming W E 1994 The New Economics (Cambridge, MA: MIT-Center for Advanced Educational Services) p 92.
[8] ICAO Cir 328 AN/190 2011 Unmanned Aircraft Systems (UAS) (International Civil Aviation Organization)
[9] Dmitriev A, Mitroshkina T and Rogachev G 2017 Structural and Parametric Analysis of Robust Design Quality of Complex Technical Systems *ITM Web Conferences* **10** 01001
[10] Dmitriev A and Mitroshkina T 2016 The ontological model and the hybrid expert system for products and processes quality identification involving the approach based on system analysis and quality function deployment *ITM Web of Conferences* **6** 02005

IOP Conf. Series: Materials Science and Engineering **476** (2019) 012010 doi:10.1088/1757-899X/476/1/012010

Influence of Inceptors on Pilot-Aircraft System Characteristics and Flying Qualities

A V Efremov, E V Efremov, Z MbiKayi and I Kh Irgaleev

Flight Dynamics and Control Department, Moscow Aviation Institute (National Research University), Moscow 125993, Russia

E-mail: pvl@mai.ru

Abstract. There are analyzed the results of experimental investigations of the feel system characteristics (inceptor's stiffness, side and central sticks, force and displacement sensing control type of inceptor) on pilot-aircraft system. The modification of the pilot structural model is considered. The mathematical modeling of pilot-aircraft system allowed to get close results to the experiments and to develop the criterion for flying qualities providing the potentiality to predict the influence of the feel system characteristics.

1. Introduction

One of the major problem in flight control system design is the determination of criteria for the flying qualities and pilot-induced oscillation prediction. The main part of them are the criteria defined in the terms of requirements to parameters of controlled element dynamics transfer function or its frequency response characteristics. Unfortunately they can not be used for selection of so-called feel system characteristics. The criteria for evaluation of these characteristics is practically absent mainly because of the limited number of researches carried out in this area. Only several of them including [1–8] were published during the last 20 – 25 years.

The results of the recent researches carried out at Moscow aviation institute and considered in this paper are the influence of some inceptor's characteristics including their type and type of signal transmitted from an inceptor to flight control system; technique and the criterion for prediction of their influence on pilot-aircraft system.

2. The experimental investigations on influence of feel system

The investigations of the stiffness was carried out for the central stick and side stick for a number of longitudinal configurations belonging to the first and third levels of flying qualities. The interval of a stiffness was 1 –30 N/sm in case of the central stick and 5 – 15 N/sm in case of the side stick. It was studies also the influence of the type of control signal transmitted to the flight control system from the inceptor. Two of them: force and displacement defining so-called force sensing control (FCS) and displacement sensing control (DCS) types of inceptors were considered below.

All parameters of feel system were simulated with help of MOOG feel system (figure 1). The experiments were carried out on MAI moving based simulator (figure 2), equipped with computer generated visual system and MOOG feel system in all control channels.

IOP Conf. Series: Materials Science and Engineering **476** (2019) 012010 doi:10.1088/1757-899X/476/1/012010

Figure 1. MOOG system for simulation of feel system parameters

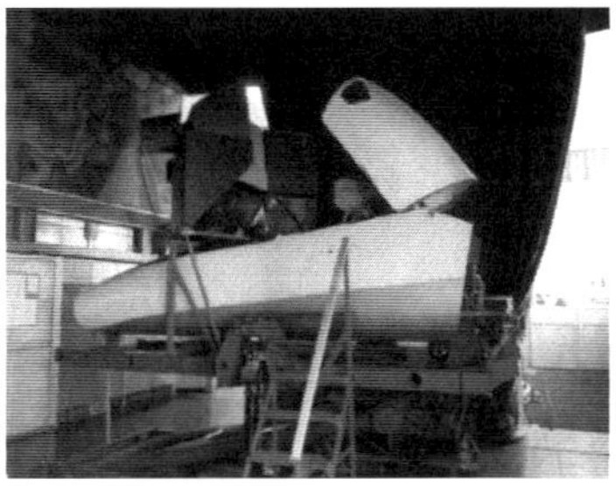

Figure 2. MAI ground based simulator

The piloting task was the single-loop tracking task. The input signal was the sum of harmonics which frequencies and amplitudes were selected to provide the power spectral density equivalent to the following equation $S_{ii} = \dfrac{K^2}{(\omega^2 + 0.5^2)^2}, \sigma_i^2 = 4\,cm^2$

The experiments carried out with the side stick for configurations HP 2.1 (I level of flying qualities) and HP 5.10 (III level of flying qualities) from Have PIO data base [9] demonstrated that in case of FSC side stick the pilot phase delay was considerably smaller. The phase delay was characterized by parameter $\Delta\varphi\big|_{\omega=10\ rad/sec} = \varphi_{FSC} - \varphi_{DSC}$, where φ_{FSC} and φ_{DSC} are the pilot phase frequency response characteristics for the FSC and DSC inceptors correspondingly defined at the frequency $\omega = 10$ rad/sec.

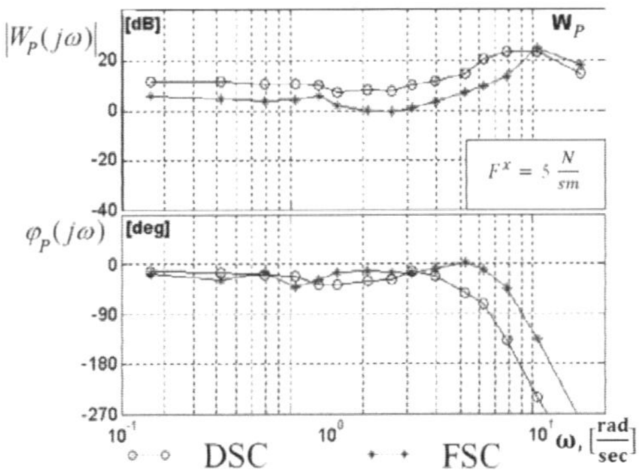

Figure 3. Pilot frequency response characteristics conf. HP 2.1

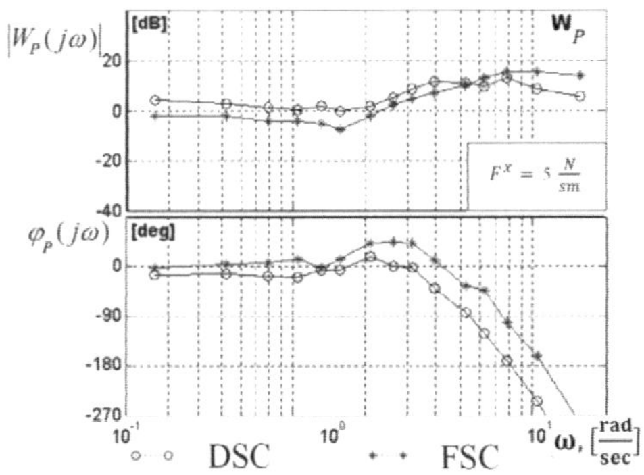

Figure 4. Pilot frequency response characteristics conf. HP 5.10

It was equal to 106 deg in the experiments with configuration HP 2.1 and the stiffness $F^X = 5\,\dfrac{N}{sm}$ (figure 3). It means that the use of FSC type of inceptor decreases the time delay close to 0.18 sec. In experiments with higher stiffness, $\Delta\varphi\big|_{\omega=10\ rad/sec}$ decreases up to 40 deg at the same frequency.

Except it, the experiments with FSC inceptor were accompanied by the decrease of variance of error σ_e^2 in 1,5 times and the increase of crossover frequency ω_C up to 30 %.

As for configuration HP5.10 the influence of the type of inceptor (DSC or FSC) on the phase delay parameter $\Delta\varphi\big|_{\omega=10\ rad/s}$ is smaller and reaches 80 deg only (figure 4).

The similar effects were observed with the central stick too. They demonstrated more sufficient and specific influence of spring stiffness on the results. For example the function $\sigma_e^2 = f(\mathrm{F}^X)$ has the minimum (see figure 5).

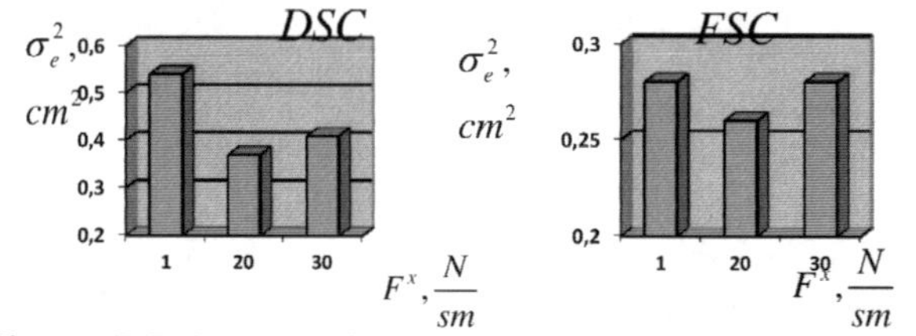

Figure 5. Influence of the stiffness on the variance of error

In case of the central stick, the difference between the pilot phase frequency characteristics in experiments with FSC and DSC type of inceptors was exposed too. The values of parameter $\Delta\varphi|_{\omega=10\ rad/sec}$ lie in the range 102 deg (conf. $\frac{K_C}{s(Ts+1)}$) ÷ 74.4 deg (conf. HP 5.10).

The influence of the central stick stiffness on pilot-aircraft system characteristics demonstrated also the decrease of the crossover frequency, resonant peak and bandwidth of closed-loop system in the case of an increase of stick stiffness.

The comparison of the accuracy achieved for the central and side sticks demonstrated higher crossover frequency and the smaller variance of error σ_e^2 in experiments with the side stick specially for the FSC type of inceptor (by 1.4 times for configuration HP-21). The same tendencies were exposed for the other dynamic configurations too.

3. Development of technique for prediction of flying qualities ensuring the selection the inceptor's characteristics

The development of criteria for FQ prediction requires the knowledge of data bases [9–11].

These data bases include 117 dynamic configurations. The experience in the use of criteria for FQ and PIO prediction demonstrates that the results of the prediction do not match often to the results of experimental investigations. One of the reasons of such disagreement is the imperfection of the data bases associated with:

- The limited number of flight tests executed for each configurations (in many cases only one test and one rating);
- Considerable variability of pilot ratings for some configurations. In some cases pilot ratings corresponded to the different FQ levels.

The averaging of such pilot ratings influences on the boundaries dividing the parameters on the ranges characterizing the specific FQ levels. It decreases the reliability of results and requires more accurate preliminary selection of the configurations for the following use of results of experiments.

Therefore it was offered to select the configurations from the data bases characterizing by more reliable in-flight results.

The "reliable" configurations were selected according to the following rules:

A. The flying qualities of each configuration had to be valuated at least two times;

B. The configurations had to be evaluated by ratings belonging to the same FQ levels;

C. The configurations related to the different levels with the difference of pilot ratings no more than one unit might be selected too.

Only 48 configurations (see table 1) from all data bases correspond to these rules. 11 configurations related to the first level of FQ, 21 configurations belong to the second level of FQ and 16 to the third level of FQ correspond to this rules.

Table 1. The configurations selected from the data bases

№	Conf.	PR	PIOR	Level	№	Conf.	PR	PIOR	Level
1	LH2.1	2; 2	1; 1	1	25	NS3E	4; 4	1.5; 1	2
2	LH4C	3; 3	1.5; 2	1	26	NS4A	5,5; 5	2.5; 2	2
3	NS1B	3,5; 3	1; 1.5	1	27	NS7G	5; 6	2; 2	2
4	NS2D	3; 2,5; 2,5	2; 1; 1	1	28	HP3.6	5; 4	2; 2	2
5	NS3C	4; 3	2; 1	1	29	NS6A	5; 6	2; 3	2
6	NS7C	3; 3; 4; 1.5	2; 2; 1; 1	1	30	NS8A	5; 4	2.5; 1	2
7	NS8C	3,5; 3	2; 1	1	31	NS6C	4; 5	2.5; 2	2
8	HP2.1	2; 2; 3	1; 1; 1	1	32	NS7E	6; 5	3; 2	2
9	HP3D	2; 2	1; 1	1	33	NS1F	8; 8	4; 4	3
10	HP4.1	3; 2; 3	1; 1; 1	1	34	NS1G	8,5; 8,5	4.5; 4	3
11	HP4.2	3; 3; 4	1; 1; 2	1	35	NS2I	8; 8	4.5; 4	3
12	LH 2A	4; 6	2; 2.5	2	36	NS4D	8; 9	3.5; 5	3
13	LH2.2	4; 4,5	2; 1	2	37	NS5D	8,5; 9; 9	4; 5; 4	3
14	LH3.0	4; 5	1; 2	2	38	NS5E	8; 8	4; 4	3
15	LH1C	4; 4	1; 1	2	39	HP2.5	10; 7; 10	4; 4; 5	3
16	LH1.1	4; 4	2; 1	2	40	HP2.8	8; 10; 8	4; 4; 4	3
17	LH2.7	7; 6	3; 3	2	41	HP3.12	7; 9	4; 5	3
18	LH3.6	7; 6	3; 3	2	42	HP3.13	10; 10	4; 5	3
19	LH4.4	7; 6	3; 3; 2	2	43	HP5.9	7; 8; 7	4; 4	3
20	NS2A	4,5; 4	2; 2	2	44	HP5.10	10; 10	5; 5	3
21	NS2H	5; 6; 5,5	2.5; 2.5; 2	2	45	LH1.3	9; 10	4; 4	3
22	NS2J	6; 6	2; 2	2	46	NS5C	9; 7	5; 5	3
23	NS3A	5; 4; 4; 4	3; 1.5; 1; 1.5	2	47	NS6E	8.5; 7	5; 4	3
24	NS3D	4; 4	2; 1	2	48	NS6F	8.5; 10; 8	4; 5; 4	3

The potentiality of the correct prediction with help of proposed rules is demonstrated below for two criteria. One of them is bandwidth-time delay criterion ($\omega_{BW} - \tau$) for prediction of flying qualities [12]. The initial and the final versions of this criterion are shown on figure 6

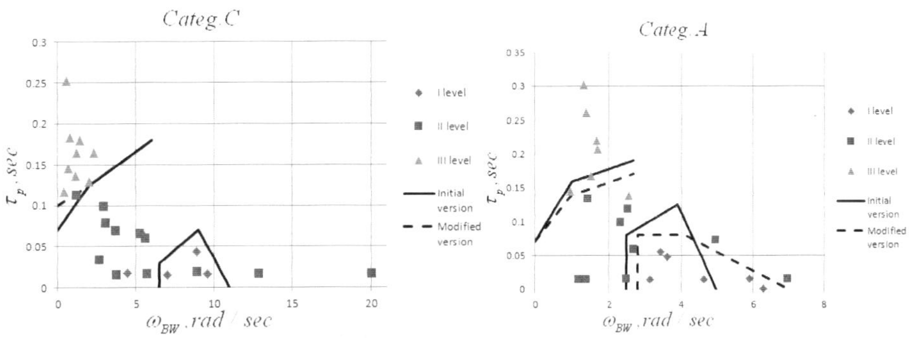

Figure 6. Criteria « $\omega_{BW} - \tau_p$ ».

The modified version allowed to increase the percentage of configurations predicted correctly from 81.3 % up to 93.8 %. Unfortunately this criterion is not sensitive to the evaluation of inceptor characteristics and it's type.

The other criterion is so-called MAI criterion [13] was defined in the terms of the resonant peak of closed-loop system r and pilot compensation parameter $\Delta\varphi$. The last one was determined as the maximum difference between pilot phase response characteristics $\varphi|_{W_c}$ and $\varphi|_{W_{c\,opt}}$ corresponding to the considered configuration W_c and optimal controlled element dynamics $W_{c_{opt}}$ in all investigated frequency range $\Delta\omega$, $\Delta\varphi = \max_{\Delta\omega}(\varphi|_{W_c} - \varphi|_{W_{c\,opt}})$.

The experiments carried out at one of MAI simulator with 66 Neal-Smith [9], Have PIO [10] and Lahos [11] configurations. The result of it is so-called MAI criterion [12] (initial version) shown on figure 7. The effectiveness of the proposed rules for selection of configurations and the following modification of criteria was investigated for the limited number of configurations (22 configurations) corresponding to this rules.

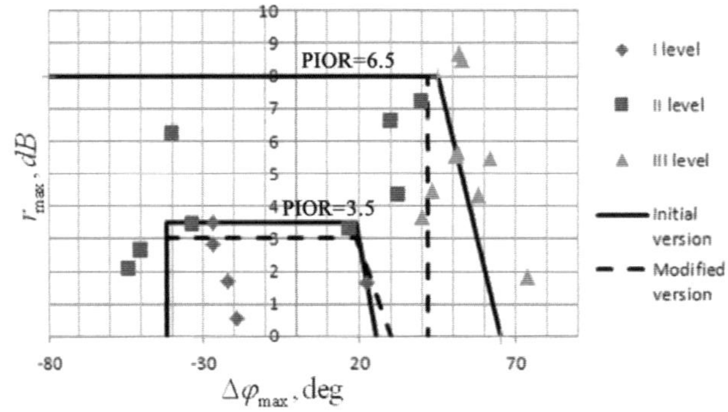

Figure 7. Modified MAI criterion with boundaries defined from experiment

The modification of the boundaries allowed to increase the number and percentage of configurations predicted correctly (up to 20 configurations and 90.9% instead of 77.2 % for the initial version of the boundaries).

Because the parameters of the criteria depend on the pilot and pilot-aircraft system parameters, which are sensitive to the different task variables, it was made the attempt to use the mathematical modeling of pilot-aircraft system with goal to calculate parameters r and $\Delta\varphi$. For that purpose the modified pilot structural model was used. The modification of this model was performed for evaluation the influence of feel system parameters on pilot-aircraft system characteristics and for selection of these parameters. The main modifications are the following:

- the separation of the limb-manipulator dynamics in two elements: dynamics of limb-muscle system and feel system dynamics;
- the introduction of additional motor noise n_c. The spectral density of this noise is $S_{n_c n_c}$ $=0.003\pi\sigma_c^2$;
- the extension of the cost function. The additional term $\beta\sigma_F^2$ was added to reflect the influence of the spring stiffness on pilot workload mentioned above and closed-loop system characteristics:

$$I = \sigma_e^2 + \alpha\sigma_c^2 + \beta\sigma_F^2 .$$

The structural pilot model is shown on figure 8.

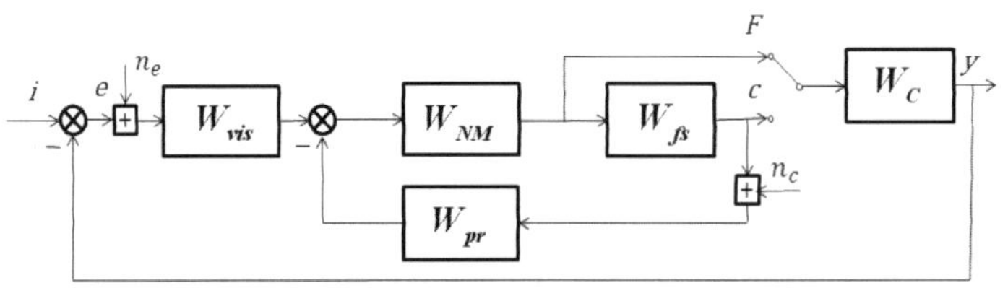

Figure 8. The structural pilot model

Here W_{fs} – the feel system dynamics;

W_{NM} – the limb muscle system dynamics $\left(W_{NM}=\dfrac{1}{T_{N*}s+1}\cdot\dfrac{e^{-\tau_N s}}{T_N^2 s^2+2T_N\xi_N s+1}\right)$;

W_{vis} – pilot error compensation dynamic $\left(W_{vis}=\dfrac{K_L(T_L s+1)}{T_I s+1}e^{-\tau s}\right)$;

W_{pr} – proprioceptive feedback dynamic $\left(W_{pr}=\dfrac{K_n s^2}{T_n^2 s^2+2T_n s+1}\right)$;

n_e – perception noise. Its power spectral density $S_{n_e n_e}=K_{n_e}\pi\dfrac{\sigma_e^2+\sigma_{\dot{e}}^2 T_L^2}{1+T_L^2\omega^2}$, K_{n_e} =0.01.

The minimization of the cost function allowed to define the parameters T_L, K_L, T_n, K_n (see the fig. 10). The other parameters of the model were constant and equal to T_N^* =0.02 sec, T_N =0.1 sec, T_I = 0.01 sec, ξ_N^* = 1.2, τ =0.2 sec, τ_N =0.08 sec. The weighting coefficient β was accepted 0.001.

The procedure for the minimization of the cost function assumes the calculation of variance $\sigma_e^2, \sigma_{\dot{e}}^2, \sigma_c^2$ from the following equations:

$$\begin{cases} A_1\sigma_e^2+B_1\sigma_{\dot{e}}^2+C_1\sigma_c^2=\sigma_{e_i}^2 \\ A_2\sigma_e^2+B_2\sigma_{\dot{e}}^2+C_2\sigma_c^2=\sigma_{\dot{e}_i}^2 \\ A_3\sigma_e^2+B_3\sigma_{\dot{e}}^2+C_3\sigma_c^2=\sigma_{c_i}^2 \end{cases}.$$

Here the symbols $\dot{e}$ and $\dot{e}_i$ are the derivatives of error. The component $\dot{e}_i$ is correlated with the input signal. The signal c is the displacement signal and c_i is its part which is correlated with the input signal.

The coefficients A_i, B_i and C_i are defined by the corresponding closed-loop transfer function, for example:

$$\begin{cases} A_1=1-\dfrac{K_{n_e}}{\pi}\displaystyle\int_0^\infty\left|\frac{W_P W_C}{1+W_P W_C}\right|^2\frac{d\omega}{1+T_L^2\omega^2} \\ A_2=-\dfrac{K_{n_e}}{\pi}\displaystyle\int_0^\infty\left|\frac{W_P W_C}{1+W_P W_C}\right|^2\frac{\omega^2}{1+T_L^2\omega^2}d\omega \\ A_3=-\dfrac{K_{n_e}}{\pi}\displaystyle\int_0^\infty\left|\frac{W_P}{1+W_P W_C}\right|^2\frac{d\omega}{1+T_L^2\omega^2} \end{cases}.$$

For FSC inceptor type $W_P=W_{vis}\dfrac{W_{NM}}{1+W_{NM}W_{pr}W_{fs}}$

For DSC inceptor type $W_P=W_{vis}\dfrac{W_{NM}W_{fs}}{1+W_{NM}W_{pr}W_{fs}}$

With help of the modified structural model the influence of stiffness and type of inceptors (FSC or DSC) was investigated. The results of the modeling demonstrated the same influence of the investigated parameters on pilot-vehicle system characteristics as in experimental investigations and a good agreement between the results of experimental and mathematical modeling (figures 9-11).

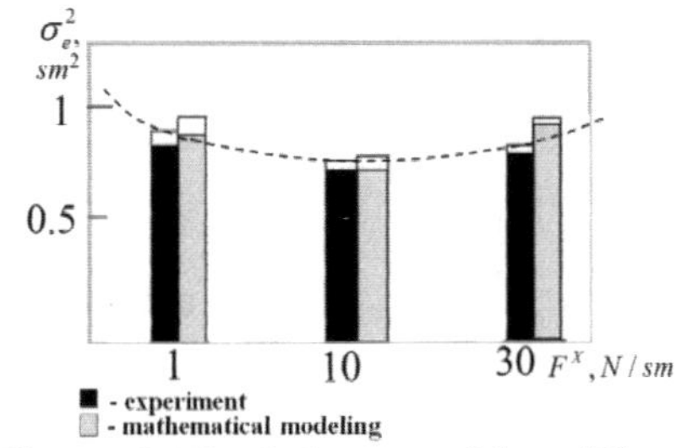

Figure 9. The influence of the stiffness

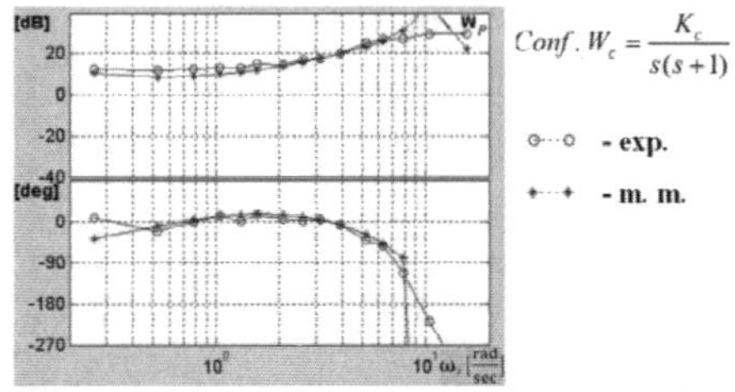

Figure 10. The agreement between the mathematical modeling and experimental results

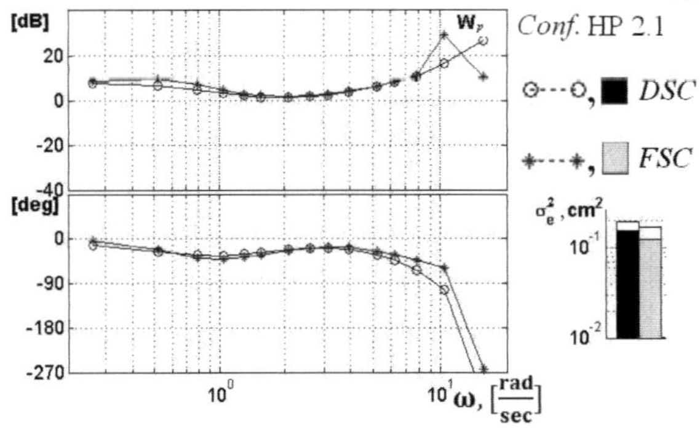

Figure 11. The influence of the type of inceptor (mathematical modeling)

The modified model allowed also to expose the difference between pilot frequency response characteristics for the different types of inceptor discussed above (figure 11).

However the attempt to use it for the calculation of resonant peak of the closed-loop system (the key parameter of MAI criterion) demonstrated bad correlation of the results with the boundaries of initial criteria. Only 52.1 % of configurations were predicted correctly (no one configuration from the first level and only 31.23% of configurations from the third level).

Because of it the attempt was undertaken to find the other parameter of closed-loop system characterizing by the better correlation with pilot rating. It was defined that such parameter is the bandwidth ω_{BW} of the close-loop system. It was accepted as the frequency corresponding to the phase of closed-loop system equal to -90 deg.

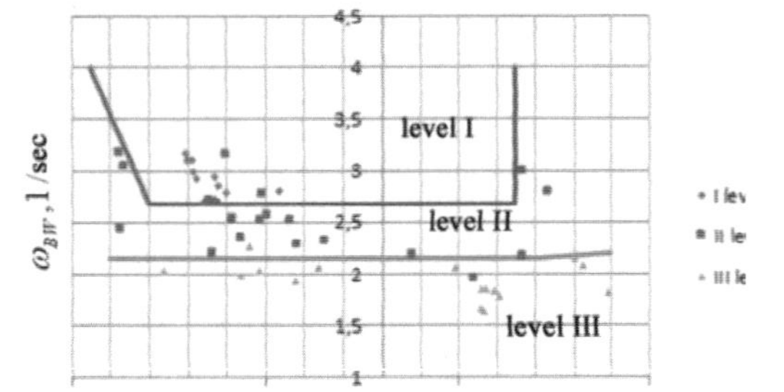

Figure 12. "New" MAI criterion

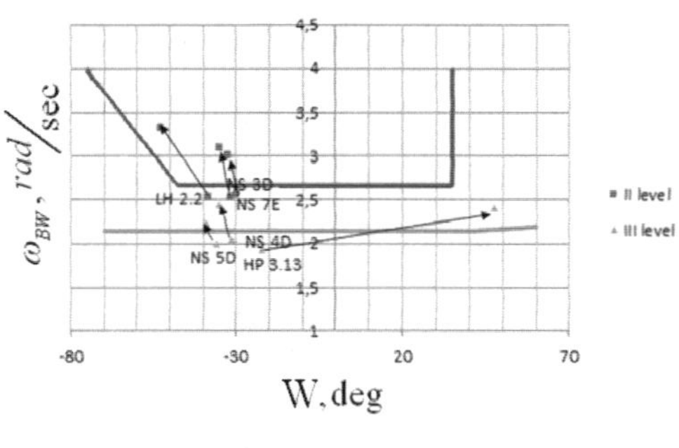

Figure 13. Evaluation of configurations with the different types of inceptor

The calculation of the parameters $\Delta\varphi$ and ω_{BW} with help of the pilot structural model allowed to get the new criterion (figure 12) characterized by high potentiality of the correct prediction.

This criterion provided 91.6 % of correct FQ prediction. It was used for the evaluation of flying qualities for two types of inceptor: DSC and FSC types.

It is seen (figure 13) that the flying qualities of configurations related to the third level (case of DSC) were improved in case when FCS type of inceptor and is used. In the last case flying qualities related to the second level of FQ. The same tendency takes place for flying qualities of the configurations related to the second level and equipped with the DCS type of inceptor. The same configurations are transformed to the first level flying qualities in case of use of FSC inceptor.

3. Conclusion

The investigation demonstrated that the use of FSC type of inceptor allows to achieve the considerable decrease of time delay in the pilot describing function (up to 0.18 sec), the improvement of the accuracy (in 1.5 times) and the increase of crossover frequency of pilot-aircraft open-loop describing function (30 % reduction). The inceptor's stiffness has a optimum value providing the minimum of error both for DSC and for FSC inceptor. Its increase causes the decrease of resonant peak of close-loop system and crossover frequency. The side stick provides higher accuracy and resonant peak in comparison with the central stick. The criteria for the prediction of flying qualities allowed to predict them with high probability and to evaluate the influence of feel system characteristics on flying qualities was developed too.

References

[1] Magdaleno R E and McRuer D T 1966 *AFFDL-TR-66-72* Effects of Manipulator Restraints on Human Operator Performance Washington D C 142

[2] Johnston D E and McRuer D T 1986 *NASA CR-3983* Investigation of Interactions Between Limb-Manipulator Dynamics and Effective Vehicle Roll Control Characteristics Washington D C 57

[3] Mitchell D G Aponso B L and Klyde D H 1992 *NASA CR-4443* Effects of Cockpit Lateral Stick Characteristics on Handling Qualities and Pilot Dynamics Washington D C 201

[4] Yilmaz D Jump M, Linghai L and Jones M 2011 *ACPO-GA-2010-266073 Deliverable No. D2.3 State-of-the-art pilot model for RPC prediction report* Aircraft and Rotorcraft Pilot Couplings – Tools and Techniques for Alleviation and Detection 35

[5] Zaychik L E Grinev K N Yashin Y P and Sorokin S A 2016 *Proceedings of the 1st IFAC Conference on Cyber-Physical & Human-System* Effect of Feel System Characteristics on Pilot Model Parameter Florianopolis 165–170

[6] Efremov A V Ogloblin A V Predtechensky A N Rodchenko V V 1992 *Pilot as a dynamic system* (Moscow: Mashinostroyeniye) (In Russian)

[7] Klyde D H and McRuer D 2009 Smart-Cue and Smart-Gain Concepts Development to Alleviate Loss of Control *Journal of Guidance, Control, and Dynamics* **32**(5) 1409–1417

[8] Klyde D H and Liang C Y 2009 Approach and Landing Flight Evaluation of Smart Cue and Smart-Gain Concepts *Journal of Guidance, Control, and Dynamics* **32**(4) 1057-1070

[9] Neal T P and Smith R E 1971 A Flying Qualities Criteria for the design of Fighter flight-control system *J of aircraft* **8**(10) 803–809

[10] Bjorkman E A et al. NT-33 1986 *USAFTPS-TR-85B-S4* Pilot induced oscillation prediction evaluation Adwards AFB 165

[11] Smith R E 1978 *AFFDL-TR-78-122* Effects of control system dynamics on Fighter approach and Landing longitudinal flying qualities Dayton **1** 35

[12] McRuer D T 1997 Aviation safety and pilot control: On the effects of aircraft pilot coupling on flight safety *National Academy Press* **189**

[13] Efremov A V Ogloblin A V and Koshelenko A V 1998 Evaluation and prediction of aircraft handling qualities Boston *Proceedings of AIAA Atmospheric Flight Mechanics Conference* 20–30

Nonparametric Method for Aircraft Flight Control

V Glasov, E Zybin and V Kosyanchuk

State Research Institute of Aviation Systems, Moscow, Russia

E-mail: vvglasov@gosniias.ru

Abstract. This paper is a part of a series of articles on nonparametric methods in dynamic systems theory. Here the authors propose a new method for inverse dynamic problem solution based only on some past measurements of controlled system input-output signals. The described nonparametric control method needs no any priori information on controlled system model parameters and does not require identification, training or statistical calculations. An example of nonparametric aircraft flight control is presented.

1. Introduction

Parametric (model-based) and nonparametric (model-free) methods are the two main parts of a complete dynamic systems control theory. The broad classification of all control methods according to the accuracy of controlled dynamic system mathematical models is presented in figure 1 [1].

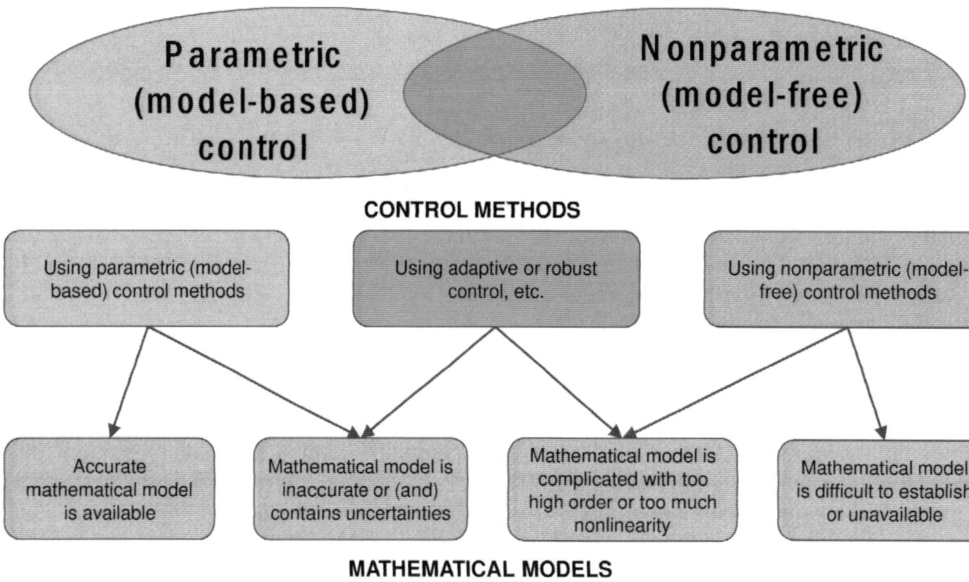

Figure 1. The classification of control methods.

Parametric control methods are the most powerful ones, but applicable only when accurate models are available or models are inaccurate and contains some uncertainties. Such mathematical models are

IOP Conf. Series: Materials Science and Engineering **476** (2019) 012011 doi:10.1088/1757-899X/476/1/012011

priori obtained from the first principles and their accurate parameters have to be calibrated on-line or off-line using measured data or estimated during identification [2–4]. Modern dynamic systems become more and more complicated. So modeling, whether by the first principles or by identification from data, is an approximation of the true system, and some error is inevitable and unmodeled dynamics always exist in the modeling and controlling processes [5]. For this reason, designing a controller using an inaccurate model with arbitrarily small errors could leads to either arbitrarily bad performance or an unstable closed-loop system [1]. Moreover, there are systems and situations when the identification problem is unsolvable in principle. Unfortunately, all the modern aircrafts belong to such systems, because their mathematical open-loop models parameters are unidentifiable in normal closed-loop operation conditions without priori information [6].

In cases when a model contains some uncertainties or inaccuracies and the required level of accuracy is higher than the classical control methods allow to achieve, the methods of adaptive or robust control are used [1, 7]. These methods are also used to control systems which models contain a large number of nonlinearities or are described by too high order complex equations. Nevertheless, adaptive control model may show unexpected behavior in the presence of certain unmodeled dynamics, and application of model-based robust control design techniques lack of adequate, practical uncertainty descriptions [1].

For controlled systems with difficult to establish or unavailable mathematical models the relatively new nonparametric control methods are used. The theory of nonparametric control is a continuation of the traditional (parametric) theory and received its development with an increase in the level of computing power of modern information technology. Nonparametric control methods can be used for systems with a well-known mathematical model and is not incompatible with parametric theory. Nevertheless, the main feature of nonparametric control theory is that it allows to solve a wide range of control tasks for any systems, including the conditions of complete lack of model parameters information. The stability and convergence of nonparametric algorithms do not depend upon the accuracy of the model, the twinborn problem of unmodeled dynamics and robustness in traditional parametric theory do not exist under nonparametric framework [1].

Although the studies on nonparametric control are still at in the inception phase, they have attracted a great deal of attentions of many researchers. At this moment a large number of nonparametric control methods are presented, which have different names depending on the group of researchers offering the method, problem statement and data usage: data-driven control (DDC), data-based control (DBC), modeless control (MLC), model-free adaptive control (MFAC), iterative feedback tuning (IFT), virtual reference feedback tuning (VRFT), iterative learning control (ILC), nonparametric model with nonparametric regulator, nonparametric dual control, nonparametric control with PID-regulator, etc. [1, 8–12]. Despite the large variety of names and approaches for solving nonparametric control problems, all these methods are united by one common condition. They find the controls based not on the known model parameters, but only on the past few measurements of its input u and output x signals (figure 2).

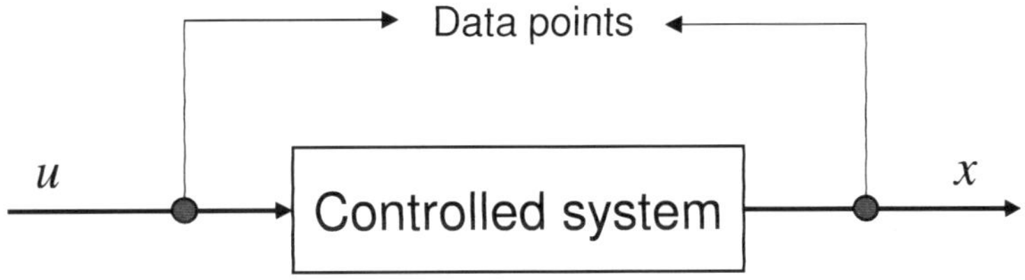

Figure 2. The general idea of nonparametric methods.

In this paper the authors propose a new method for nonparametric control problem solution, significantly expanding the applicability of nonparametric dynamic systems theory [13–16].

IOP Conf. Series: Materials Science and Engineering **476** (2019) 012011 doi:10.1088/1757-899X/476/1/012011

2. Nonparametric aircraft flight control problem formulation and solution

Let's an aircraft dynamics is described by a discrete state space model in the form the following parametric vector

$$x_{i+1} = Ax_i + Bu_i$$

or matrix

$$\begin{bmatrix} x_{i-h} & \dots x_i & x_{i+1} \end{bmatrix} = A\begin{bmatrix} x_{i-h-1} & \dots x_{i-1} & x_i \end{bmatrix} + B\begin{bmatrix} u_{i-h-1} & \dots u_{i-1} & u_i \end{bmatrix}, \quad (1)$$

expressions, where *A*, *B* – eigen-dynamics and control efficiency matrices, *x* – state vector, *u* – control vector, *h* – number of observations, *i* – discrete time.

The scheme of the aircraft dynamic model is presented in figure 3.

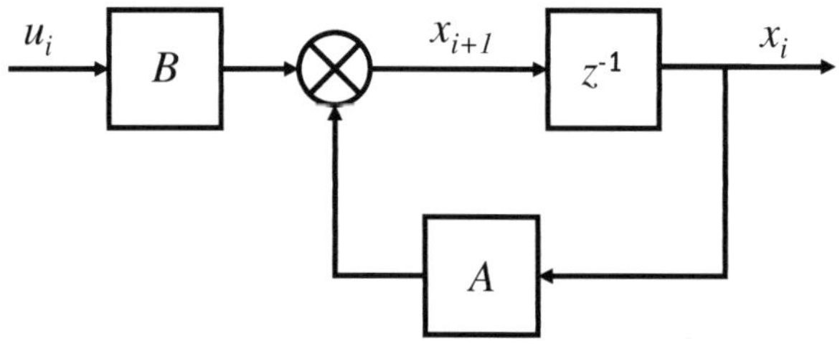

Figure 3. The scheme of a discrete aircraft dynamic model.

The problem of nonparametric control of the aircraft, described by the equation (1), is to find the controls u_i, knowing only the desired states x_i, without any priori information on the aircraft model parameters *A*, *B*.

To solve this problem, let's write the equation (1) in a block-matrix form

$$\begin{bmatrix} X_i & x_{i+1} \end{bmatrix} = A\begin{bmatrix} X_{i-1} & x_i \end{bmatrix} + B\begin{bmatrix} U_{i-1} & u_i \end{bmatrix}, \quad (2)$$

where $X_i = \begin{bmatrix} x_{i-h} & \dots x_i \end{bmatrix}$, $X_{i-1} = \begin{bmatrix} x_{i-h-1} & \dots x_{i-1} \end{bmatrix}$, $U_{i-1} = \begin{bmatrix} u_{i-h-1} & \dots u_{i-1} \end{bmatrix}$, and convert it to the following linear matrix equation

$$\begin{bmatrix} I - A \end{bmatrix}\begin{bmatrix} X_i & x_{i+1} \\ X_{i-1} & x_i \end{bmatrix} = B\begin{bmatrix} U_{i-1} & u_i \end{bmatrix}, \quad (3)$$

representing the aircraft eigen-dynamics parameters identification problem.

Equation (3) has the form of a right-hand side matrix equation

$$ZC = D,$$

which is solvable if and only if

$$D\overline{C^R} = 0, \quad (4)$$

where $\overline{C^R}$ – the right-hand full rank zero divisor of the matrix *C*, such that $C\overline{C^R} = 0$, which formally describes all the linear dependent columns of matrix *C* [6, 13–16].

According to (4), we always can find such *h*, that the necessary and sufficient solvability condition of (3) has the form of equality

$$B\begin{bmatrix} U_{i-1} & u_i \end{bmatrix}\overline{\begin{bmatrix} X_i & x_{i+1} \\ X_{i-1} & x_i \end{bmatrix}^R} = 0, \quad (5)$$

which can be always achieved also by fulfilling the sufficient condition

$$\begin{bmatrix} U_{i-1} & u_i \end{bmatrix}\overline{\begin{bmatrix} X_i & x_{i+1} \\ X_{i-1} & x_i \end{bmatrix}^R} = \begin{bmatrix} U_{i-1} & u_i \end{bmatrix}\begin{bmatrix} r_i \\ 1 \end{bmatrix} = 0, \quad (6)$$

where

$$\begin{bmatrix} r_i \\ 1 \end{bmatrix} = \overline{\begin{bmatrix} X_i & x_{i+1} \\ X_{i-1} & x_i \end{bmatrix}^R} \quad (7)$$

is the single-column normalized right-hand zero divisor, such that

$$\begin{bmatrix} X_i & x_{i+1} \\ X_{i-1} & x_i \end{bmatrix}\begin{bmatrix} r_i \\ 1 \end{bmatrix} = 0. \tag{8}$$

So, from (6) we can find the nonparametric aircraft control algorithm of the form

$$u_i = -U_{i-1} r_i , \tag{9}$$

allowing obtaining the required controls u_i based only on some h past measurements and the desired state x_{i+1} without any information on the model parameters A and B.

3. Nonparametric aircraft flight control example

To verify the proposed nonparametric control method let's simulate a midrange passenger aircraft dynamics by discrete state-space model (1), where $x=[\Delta V\ \ \Delta\alpha\ \ \Delta\vartheta\ \ \Delta\omega_z\ \ \Delta\beta\ \ \Delta\omega_x\ \ \Delta\omega_y\ \ \Delta\gamma]^T$; V – flying speed (km/h); $\alpha,\vartheta,\beta,\gamma$ – angles of attack, pitch, slip, and roll (deg); $\omega_x,\omega_y,\omega_z$ – angular rates of roll, yaw, and pitch (deg/s); $u=[u_{e\_l}\ \ u_{e\_r}\ \ u_{st}\ \ u_{ru}\ \ u_{a\_l}\ \ u_{a\_r}\ \ u_{i\_l}\ \ u_{i\_r}]^T$; $u_{e\_l},u_{e\_r},u_{st},u_{ru},u_{a\_l},u_{a\_r},u_{i\_l},u_{i\_r}$ – deflection angles of right and left elevators, stabilizer, rudder, right and left ailerons, right and left interceptors (deg).

$$A = \begin{bmatrix} 0.99770 & -1.02099 & -0.98098 & 0 & 0 & 0 & 0 & 0 \\ -0.00057 & 0.90700 & 0.00008 & 0.10000 & 0 & 0 & 0 & 0 \\ 0.00057 & 0.09300 & 0.99992 & 0 & 0 & 0 & 0 & 0 \\ -0.00002 & -0.07897 & -0.00001 & 0.97279 & 0 & 0 & 0 & 0 \\ 0 & 0 & 0 & 0 & 1.00278 & 0.01139 & 0.09876 & 0.01370 \\ 0 & 0 & 0 & 0 & -0.34455 & 0.65043 & -0.08993 & 0 \\ 0 & 0 & 0 & 0 & -0.19870 & 0.02042 & 0.89593 & 0 \\ 0 & 0 & 0 & 0 & 0 & 0.10000 & -0.01087 & 1 \end{bmatrix};$$

$$B = \begin{bmatrix} 0 & 0 & 0 & 0 & 0 & 0 & 0.00052 & 0.00052 \\ 0.00178 & 0.00178 & -0.00084 & 0 & -0.00010 & -0.00010 & -0.00070 & -0.00070 \\ -0.00178 & -0.00178 & 0.00084 & 0 & 0.00010 & 0.00010 & 0.00070 & 0.00070 \\ -0.03378 & -0.03378 & -0.13433 & 0 & -0.00501 & -0.00501 & 0.01271 & 0.01271 \\ 0 & 0 & 0 & -0.00054 & 0 & 0 & -0.00022 & 0.00022 \\ 0.02052 & -0.02052 & 0 & -0.01973 & 0.05419 & -0.05419 & -0.16357 & 0.16357 \\ -0.00092 & 0.00092 & 0 & -0.31463 & -0.00579 & 0.00579 & 0.02973 & -0.02973 \\ 0 & 0 & 0 & 0 & 0 & 0 & 0 & 0 \end{bmatrix}.$$

Figure 4 shows the scheme of the dynamic aircraft model with a nonparametric control device.

Figure 5 depicts the parallel simulation results of parametric (blue line) and nonparametric (red line) control algorithms with simulation time of 5 seconds and sample rate $\Delta t = 0.1$ seconds. Note that the nonparametric controls appear only at the 14th simulation step. This is due to the fact that in the current situation for the existence of the right-hand zero divisor (8) it is necessary to accumulate 13 columns of past states and controls to detect their linear dependences. So, the sample width $h=13$, and the nonparametric algorithm tune time $t_{tune} = \Delta t(h+1) = 1.4\,\text{sec}$. The simulation results show that nonparametric controls completely coincide with reference ones after tune time.

IOP Conf. Series: Materials Science and Engineering **476** (2019) 012011 doi:10.1088/1757-899X/476/1/012011

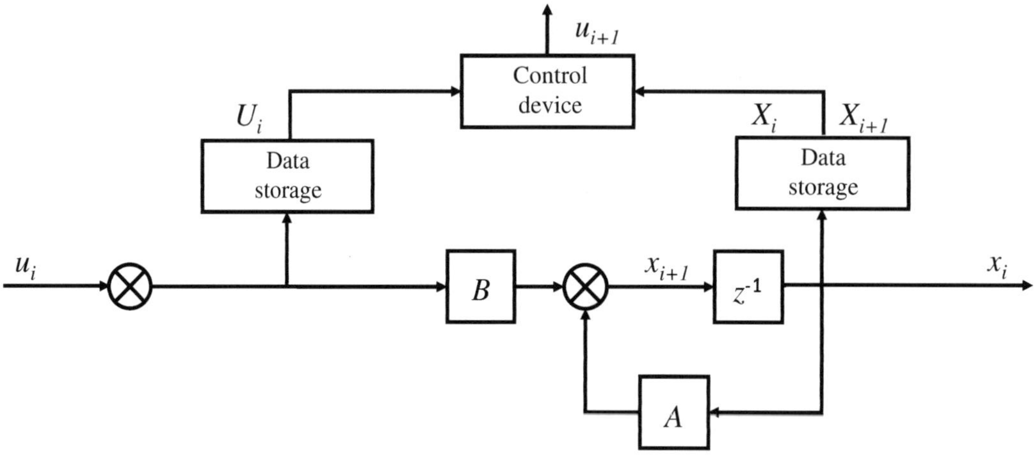

Figure 4. The scheme of aircraft discrete dynamic model with a nonparametric control.

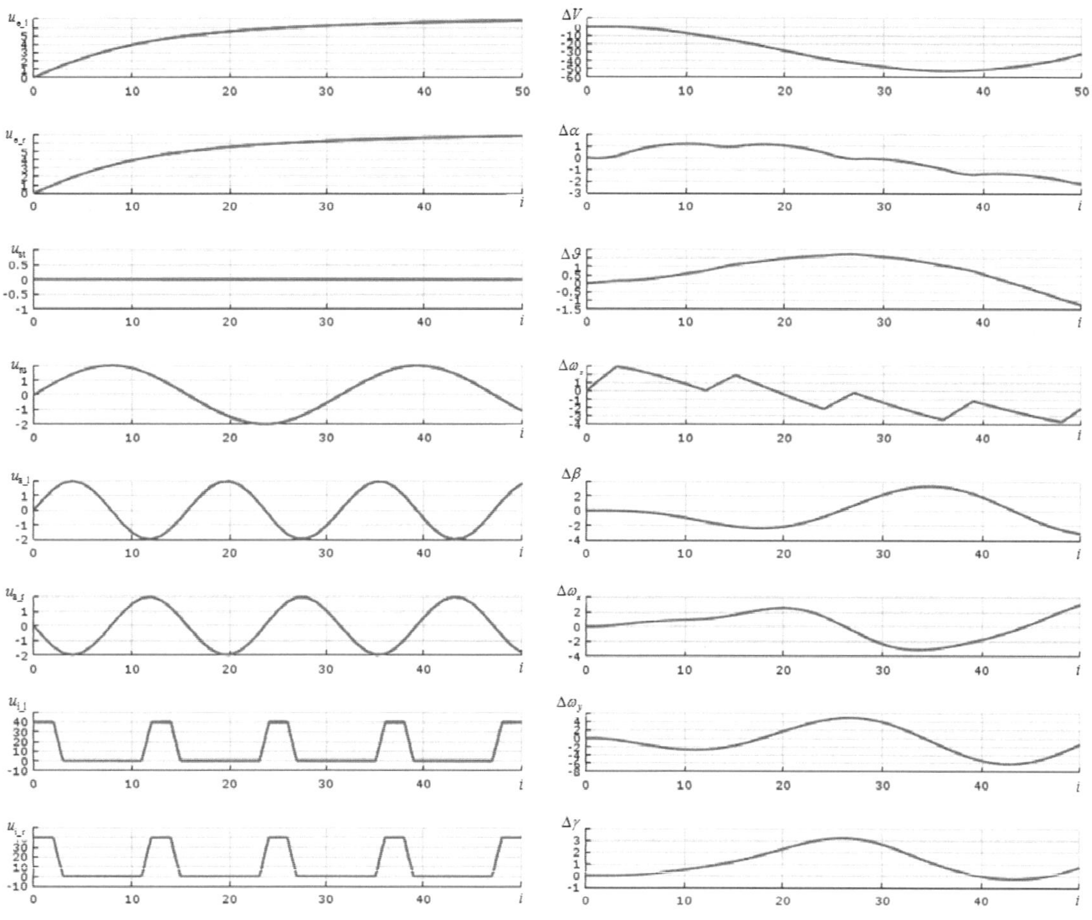

Figure 5. Aircraft's controls and states.

Acknowledgments
This work was supported by Russian Foundation for Basic Research (RFBR), projects 17-08-01445a, 18-08-00453a.

References

[1] Hou Z S and Wang Z 2013 *Information Sciences* **235** 3-35
[2] Korsun O N, Stulovskii A V, Balyk O A and Zolotaeva M V 2018 *IOP Conference Series: Materials Science and Engineering* **312** 012015
[3] Korsun O N, Stulovskii A V, Ovcharenko V N and Kanyshev A V 2018 *Journal of Computer and Systems Sciences International* **57** (3) 374-389
[4] Korsun O N, Om M H, Latt K Z and Stulovskii A V 2017 *Procedia Computer Science* 67-74
[5] Korsun O N, Stulovskiy A V and Bukovskiy G A *Herald of computer and information technologies.* 2018 **6** (168) 27-36 (in Russian)
[6] Zybin E Yu 2015 *Izvestiya SFedU. Engineering Sciences* **4** (166) 160-170 (in Russian)
[7] Evdokimenkov V N, Krasil'shchikov M N, Om M H, Korsun O N and Sebryakov G G 2017 *Procedia Computer Science* 126-134
[8] Zybin E Y and Kos'yanchuk V V 2016 *Journal of Computer and Systems Sciences International* **55** (4) 546-557
[9] Zybin E, Kosyanchuk V and Karpenko S 2017 *MATEC Web of Conferences* **99** 03011
[10] Karpenko S S, Zybin E Yu and Kosyanchuk V V 2018 *IOP Conference Series: Materials Science and Engineering* **312** 012010
[11] Terent'yev M N, Karpenko S S, Zybin E Yu and Kosyanchuk V V 2018 *IOP Conference Series: Materials Science and Engineering* **312** 012025
[12] Hou Z, Chi R and Gao H 2017 *IEEE Transactions on Industrial Electronics* **64** (5) 4076-4090
[13] Yin S, Li X, Gao H and Kaynak O 2015 *IEEE Transactions on Industrial Electronics* **62** (1) 657-667
[14] Hou Z and Jin S 2013 *Model free adaptive control: theory and applications* (NW: CRC press)
[15] Campestrini L, Eckhard D, Bazanella A S and Gevers M 2017 *Journal of the Franklin Institute* **354** (6) 2628-2647
[16] Hou Z, Gao H and Lewis F L 2017 *IEEE Transactions on Industrial Electronics* **64** (5) 4070-4075

IOP Conf. Series: Materials Science and Engineering **476** (2019) 012012 doi:10.1088/1757-899X/476/1/012012

Flight Simulator with a Wake Vortex Regime

I G Golovnev[1], V V Vyshinsky[2], K V Lapshin[1], R Yu Zimin[1] and D S Lyubaeva[3]

[1]Federal state unitary enterprise «State Research Institute of Aviation Systems» Moscow, 125319, Russia
[2]Federal state unitary enterprise «Central Aerohydrodynamic Institute», Zhukovsky 140180, Russia
[3]National Research University Higher School of Economics, Moscow 101000, Russia

E-mail: golovnev@gosniias.ru

Abstract. Flight simulator with imitation vortex trail - imitation of hitting a vortex wake on an aircraft simulator made in accordance with Doc 9625 [1] is proposed to be performed using a model of an onboard warning system [2] of an aircraft entering a vortex wake from another aircraft. Through the example of an airplane entering an A-380 wake vortex the paper explains the technology of simulation of an airplane upset and recovery.

1. Introduction

Flight simulator with imitation vortex trail - imitation of hitting a vortex wake on an aircraft simulator made in accordance with Doc 9625 [1] is proposed to be performed using a model of an onboard warning system [2] of an aircraft entering a vortex wake from another aircraft.

In the well-known aircraft simulators [3–5] models of the vortex wake are used, based on the calculation of its characteristics and visualization, but training on such a simulator does not develop skills for piloting in a vortex wake.

In this article, in order to provide piloting training in the conditions of a vortex wake, it is proposed to introduce into the simulator a block of imitation of hitting the vortex wake based on the simulation of an onboard vortex wake early warning system [2]. In this case, the "integration" of the mathematical model of the vortex wake into the mathematical model of the aircraft is carried out, as well as the visualization and/or indication of the vortex wake. As a result the simulator enables the pilot's training in the vortex wake conditions.

2. Description of the research

Flight simulator with imitation vortex trail - imitation of hitting a vortex wake on an aircraft simulator made in accordance with Doc 9625 is proposed to be performed using a model of an onboard warning system of an aircraft entering a vortex wake from another aircraft [2, 6] based on measuring the pressure in a vortex on board the aircraft. The number and location of pressure receivers are determined for each aircraft.

The onboard vortex wake early warning system transmits the measured air presser to the computer , where the assessment of the degree of impact of the vortex wake on the aircraft is made - the change in the coefficients of roll moment, drag C_x and lift C_y forces The obtained results are transmitted to the indicator and to the control system. The efficiency of this onboard vortex early warning system is estimated using the example of a crash of an DR 400 in the vortex wake of an AN-2 aircraft [7].

The calculations in [7] made it possible to estimate the vortex for AN-2 - circulation and maximum tangential velocity in the vortex during takeoff (Figure 1-2).

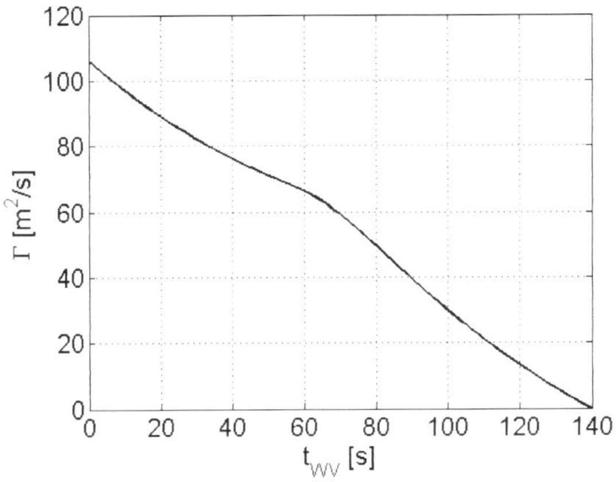

Figure 1. Graph of distribution of circulation

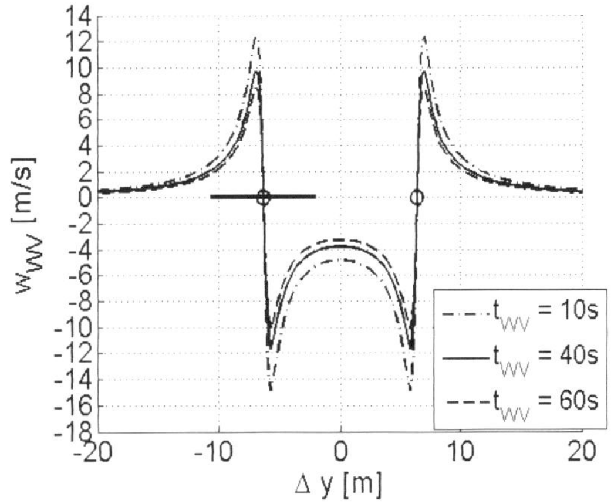

Figure 2. Graph of distribution tangential velocity

The DR 400 catastrophe occurred due to its falling into the vortex wake at 40 sec after its formation, with the maximum tangential velocity in the vortex being $V_\tau \approx 12$ m/s. Using these data, using the procedure described in [2], we calculated vacuum in the vortex for AN-2 during takeoff (figure 3). At the time of the crash, the vacuum in the vortex was 100 Pa. If an early warning system would had been on board, the pilot could had information about a vortex hit in at least 20 sec before the disaster, since after 20 sec the required measurement accuracy is obtained.

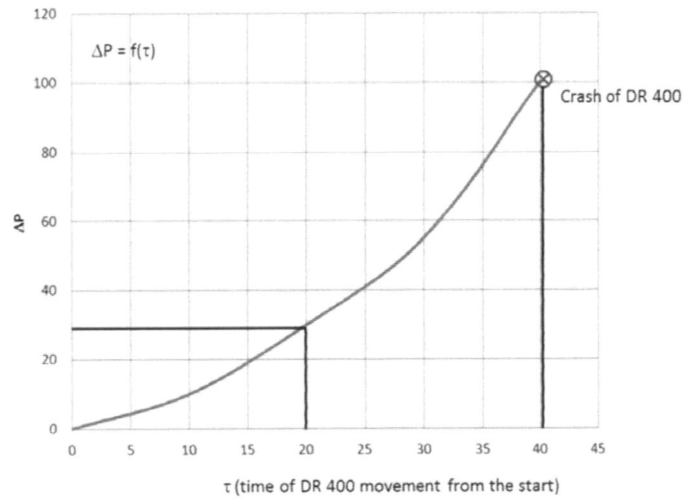

Figure 3. The vacuum in the vortex for the AN-2

The initial data for simulating an airplane into a vortex wake from another airplane are: atmospheric characteristics, characteristics of a vortex generator airplane, characteristics of an airplane which follows, flight mode scenario. The program contains a program for calculating the characteristics of the vortex: the vacuum in the vortex core, the maximum tangential velocity, the increments of the coefficients of the aerodynamic forces C_x, C_y and roll moment M_x. The calculation is performed in accordance with the recommendations [8]. The aerodynamic coefficients are fed into a mathematical model, which provides the level of allowable changes in the characteristics of the aircraft due to the influence of the vortex wake.

The simulation technology on the aircraft simulator is considered on the example of an airplane with a wing span of 40 m which hits in the swirl trail from the A-380. Figure 4 presents the results of calculating the maximum tangential velocity in a vortex at different distances [2]. In this case the change of the coefficient of the moment of roll due to the vortex wake, as the most critical for this situation, was taken as a criterion. The aircraft was located in a vortex wake at various distances from the A-380 aircraft. For these distances, calculations of the increment of the momentum coefficient of the roll is presented in figure 5. The figure shows that when the aircraft enters the center of the A-380 wing vortex, there are moments of roll that cannot be compensated with the ailerons. Fgure 6 shows a graph of the change in the maximum tangential velocity in the vortex from A-380 depending on the distance X. The resulting roll moment is not compensated by ailerons at a tangential speed of 18 m/s (X = 8 km). The red horizontal line is the boundary above which the tangential velocities of the vortex will be dangerous for the aircraft [2].

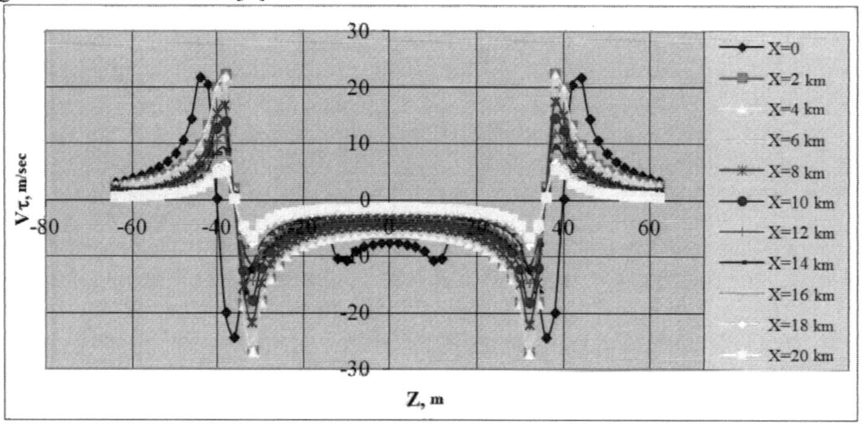

Figure 4. Distribution of maximum tangential velocity

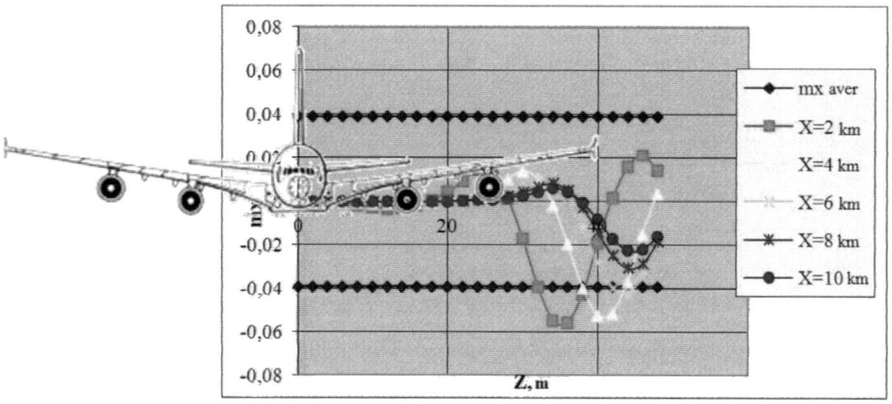

Figure 5. Distribution of the roll moment

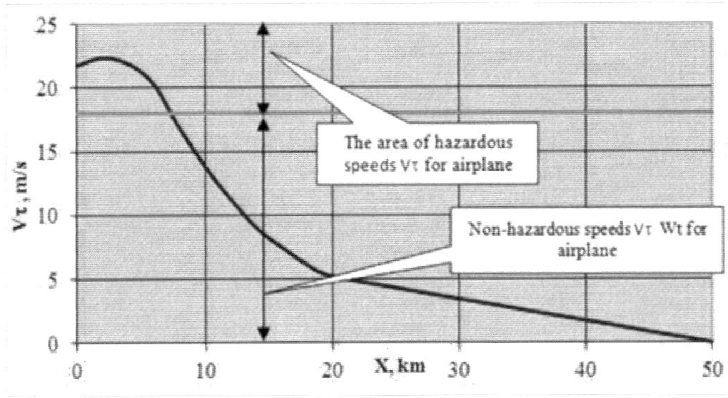

Figure 6. Graph of change of maximum tangential velocity in a wake vortex

The figure 7 shows changes $\Delta P_{v_{\tau max}}$ in the vortex wake behind A-380 - acceptable critical vacuum $\Delta P_{v_{\tau max}}(add) = 66.7$ Pa for X = 8 km. Let us assume that the warning to the pilot of hitting the vortex wake begins with ΔP = 8 Pa (distance A-380 20 km, figure 7). Since the maximum permissible calculation conditions is reached at X = 8 km, there is enough time to decide if the change in the flight mode to ensure the aircraft's vortex safety is needed.

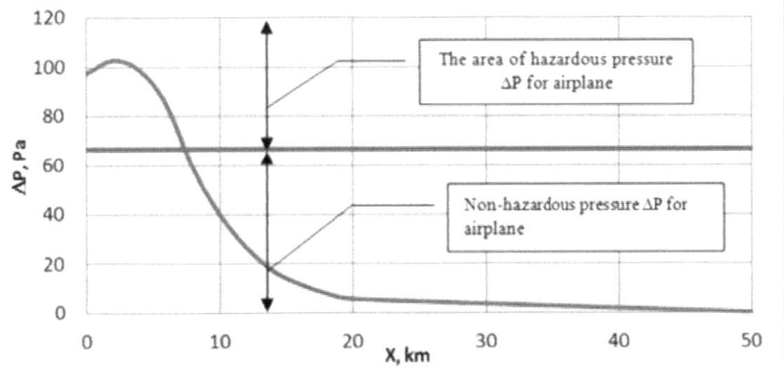

Figure 7. Graph of pressure change in the vortex

The above simulation technology for hitting a vortex wake on an aviation simulator (integrating a mathematical model of an early warning system into a mathematical model of an aircraft) allows the pilot to evaluate the influence of the vortex wake and make a decision to change the flight mode based on the information about the intensity of the vortex wake online mode.

To increase efficiency of an aircraft simulator, it is advisable to include into the process of imitation of a vortex wake its visualization on the pilot's display according to the results of calculating the parameters of the vortex of the airplane generator to the airplane follower (figure 8).

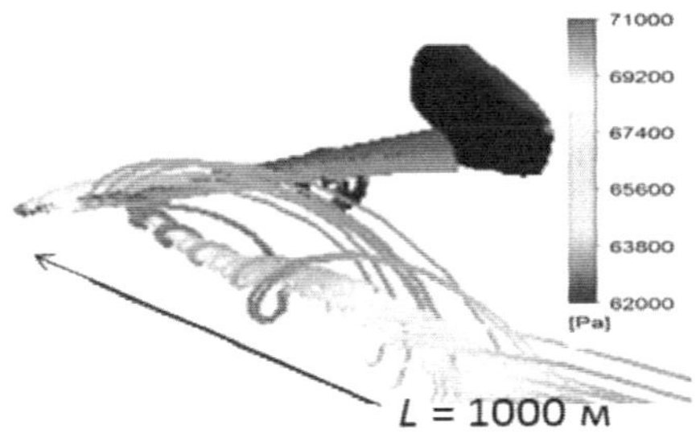

Figure 8. Results of mathematical modeling

3. Conclusion

The developed technology allows simulation of wake vortex on an aircraft simulator, based on the simulation of an onboard early warning system of pilot about entering wake vortices from another aircraft.

References

[1] Doc 9625 2015 *Guidance on the criteria for the qualification assessment of flight simulators* (ICAO)
[2] Golovnev I, Vyshinsky V, Zhelannikov A and Lapshin K 2018 Design concepts of an onboard early warning system of pilot about entering wake vortices from another aircraft *Civil aviation high technologies (Scientific Bulletin)* **21** 84-95
[3] Baranov N and etc. 2008 *Aviation Simulator* (RF Patent No. 2324982) (in Russian)
[4] Animitsa O, Gaifullin A, Ryzhov A and Sviridenko Yu 2015 *Modeling at the flight stand of aircraft refueling in flight* vol 7 (Moscow: MIPT press) (in Russian)
[5] Vyshinski V, Ivanov K, Terpugov A 2015 *Simulation of complex flight regimes on flight stands, taking into account atmospheric turbulence* (Moscow: MIPT press, vol 7) pp 36 – 43
[6] Golovnev I. and Lapshin K 2017 The on-board measurement system of air data as a basis for the airplane to be notified about its entry into a vortex formation from another airplane *All-Russian Scientific and Technical Conference "Navigation, Aiming and Control of Aircraft press* pp 218-220 (in Russian)
[7] German Aerospace Center (DLR) 2014 *Wake turbulence - DLR tests warning system in flight experiments* (Electronic materials) https://www.dlr.de/dlr/en/desktopdefault.aspx/tabid-10261/371_read-9997//usetemplate-print/
[8] Technical Report 2011 *Technical Report to Support the Safety Case for the Recategorization of ICAO Wake Turbulence Standards Methodology For Re-Categorization of ICAO. Wake Turbulence Standard Produced for the ICAO. Wake Vortex Task Force (version 0.1)* (ICAO)

IOP Conf. Series: Materials Science and Engineering **476** (2019) 012013 doi:10.1088/1757-899X/476/1/012013

Airship Balancing and Stability at Longitudinal Established Motion

V T Grumondz

Moscow Aviation University, Moscow, 125080, Russia

E-mail: V.grumondz @gmail.com

Abstract. Dynamical differential equations of the airship motion as well as mathematical model of the longitudinal steady motion of an airship are designed. The conditions for longitudinal stability of horizontal motion are obtained. The values of the speed at which the established motion of the airship is impossible are found.

1. Introduction

Recently there is an increasing interest towards aerostatic aircraft, including manned and unmanned airships [1–27]. Airship flight dynamics is significantly different from airplane [3-7], and has not been deeply studied so far [11-16]. This article is dedicated to analysis of airship longitudinal balance at established motion trajectory.

Airship dynamics is somehow similar to a similar shape underwater vehicle dynamics. This is mostly due to a great influence of the aerostatic and hydrodynamic forces and their moments, as well as added masses, and inertial moments [21-25]. This is also due to the fact that airship weighted density is close to ambience density (which is also true for submerged vehicle). Due to that it is logical to use the mathematical model of the motion of a fully submerged vehicle for the equations describing an airship motion.

2. Longitudinal established motion

Let's assume that the airship has the classical shape (body of rotation) with cruciform tail, gondola, and two symmetrically located propeller engines. We'll also assume that the airship is solid (non-deformable) and the variations of masses, insertion moments and added masses are insignificant ($m = \text{const}$, $J = \text{const}$, $\lambda_{ik} = \text{const}$). Airship center (or airship pole) is the origin of the coordinate system $Oxyz$. Coordinates of the center of masses in $Oxyz$ will be x_c, y_c, z_c, and we'll assume that $z_c = 0$. We will also consider the centrifugal inertion moments J_{xz}, J_{yz} as minor.

Dynamical equations of the airship motion result from general mechanics theorems describing quantity of motion an momentum of quantity of motion of the mechanical system. In vectoral form those equations will be:

$$\frac{d\vec{Q}}{dt} + \vec{\omega} \times \vec{Q} = \vec{F}\,; \qquad \frac{d\vec{L}}{dt} + \vec{\omega} \times \vec{L} + \vec{V} \times \vec{Q} = \vec{M}\ ,$$

where $\vec{Q},\vec{L}$ – are the main vector and main momentum of motion of the body (airship or underwater vehicle) in ideal liquid (or air around the airship); ω - angular velocity; $\vec{F}$, $\vec{M}$ – main vector and main moment of external forces acting on the airship, and excluding inertion forces. Equation projections on axes Ox, Oy and momentum equation on axis Oz will be

$$\frac{d}{dt}\frac{\partial T^*}{\partial V_x}+\omega_y\frac{\partial T^*}{\partial V_z}-\omega_z\frac{\partial T^*}{\partial V_y}=F_x;$$

$$\frac{d}{dt}\frac{\partial T^*}{\partial V_y}+\omega_z\frac{\partial T^*}{\partial V_x}-\omega_x\frac{\partial T^*}{\partial V_z}=F_y;$$

$$\frac{d}{dt}\frac{\partial T^*}{\partial \omega_z}+\omega_x\frac{\partial T^*}{\partial \omega_y}-\omega_y\frac{\partial T^*}{\partial \omega_x}+V_x\frac{\partial T^*}{\partial V_y}-V_y\frac{\partial T^*}{\partial V_x}=M_z$$

Right part of this equation is the vector projection

$$\bar{F}=\bar{G}+\bar{A}+\bar{P}+\bar{R},\ \bar{M}=\bar{M}_G+\bar{M}_P+\bar{M}_R,$$

where $\bar{G}$ is the airship weight, including weight of the air and gas inside the airship; $\bar{A}$ is the aerostatic force vector; $\bar{P}$ is the vector of engines forces; $\bar{R}$ is the vector of aerodynamic forces; $\bar{M}_G,\bar{M}_P,\bar{M}_R$ are the main momentums of gravity, engines propulsion, and aerodynamic forces in relation to the pole O. For vectors $\bar{R}$, $\bar{M}$ projections we will use the typical aerodynamic coefficients c_x,c_y,c_z,m_x,m_y,m_z.

For classic shape airships longitudinal dimension is $l=U^{1/3}$, and typical area is $S=U^{2/3}$.

We will also assume that the area of gondola and engines are minor in relation to the airship area, so the axis Ox will be the axis of geometrical symmetry of the whole airship. Then only the values of the following added masses $\lambda_{ii}, i=1...6, \lambda_{26}, \lambda_{35}$ will not equal to zero, so the equation for the kinetic energy of the "ideal liquid" is

$$T=\frac{1}{2}\left(\lambda_{11}V_x^2+\lambda_{22}V_y^2+\lambda_{66}\omega_z^2\right)+\lambda_{26}V_y\omega_z.$$

Certain kinematic and geometric equations can be added to the above equation, but those are not presented in this article.

The equations for the longitudinal established motion of the airship are

$$(\rho gU-G)\alpha+P\cos\varphi=c_{x_0}qS-(\rho gU-G)\Theta;$$

$$c_y^{\alpha}qS\alpha+c_y^{\delta}qS\delta+P\sin\varphi=-(\rho gU-G);$$

$$\left(Gy_c+m_z^{\alpha}qU\right)\alpha+m_z^{\delta}qU\delta+P(x_P\sin\varphi-y_P\cos\varphi)=G(x_c-y_c\Theta),$$

where $c_{x_0}, c_y^{\alpha}, c_y^{\delta}, m_z^{\alpha}, m_z^{\delta}$ – aerodynamic coefficients.

In order to evaluate the flight characteristics and controllability of the airship we need first to define the values of required propulsion P and balancing angles α,δ_B, which ensure motion alongside a linear trajection with given velocity V_* and angle ϑ_*. In order to get quantitative parameters we shall investigate a classic composition airship; equation for contours will be

$$\bar{y}=0,972\sqrt{\bar{x}(1-\bar{x})(1,5-\bar{x})};$$

$$\bar{x}=\frac{x}{L},\ \bar{y}=\frac{y}{D},$$

L – length and D – maximum diameter of the airship. As base scenario for the calculation let us assume the area of the airship is $U=550m^3$; length $L=28,18$m; maximum diameter $D=6,26$m; ratio of airship length to diameter is $\lambda=L/D=4,5$; midship area $S_M=\pi D^2/4=30,78m^2$ $S_M=\pi D^2/4=30,78m^2$; characteristical area is $S=U^{2/3}=67,08m^2$; characteristical longitudinal

IOP Conf. Series: Materials Science and Engineering **476** (2019) 012013 doi:10.1088/1757-899X/476/1/012013

size is $l = U^{1/3} = 8{,}19m^2$; distance from the center to the nose is 12,68m; relative area of each stabilizer is $\bar{S}_{st} = S_{st}/U^{2/3} = 0{,}085$; ratio of each stabilizer length to its chord is $\lambda_{st} = 0{,}82$; relative distance from stabilizers center to airship center is $\bar{l}_{st} = l_{st}/U^{1/3} = 1{,}39$; relative area of each rudder is $\bar{S}_r = S_r/U^{1/3} = 0{,}026$; ratio of each rudder length to its chord is $\lambda_r = 2{,}6$; relative distance from the rudders axis to the center of the airship is $\bar{l}_P = l_P/U^{1/3} = 1{,}55$; range of horizontal and vertical rudders angles is $\pm 30^{\circ}$; ratio of airship size to the size of a cylinder with same dimensions is $\psi_U = U/LS_M = 0{,}635$.

At cruising mode with permanent altitude and speed $\Theta = 0, \varphi = 0, x_c = 0$, $V = \text{const}$ we can define the dependence of the balancing angles of attack and elevating rudders range from the speed of established horizontal motion (fig. 1,2, where $\varPhi = (\rho g U - G)$).

Figures 1 and 2 show the existing of a V-values range for which the balancing of the airship is absent.

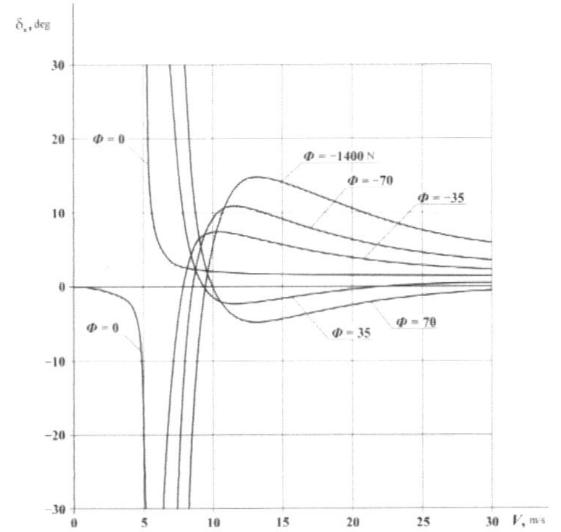

Figure 1. Dependence of the balancing angles of elevating rudders range from the speed of established horizontal motion

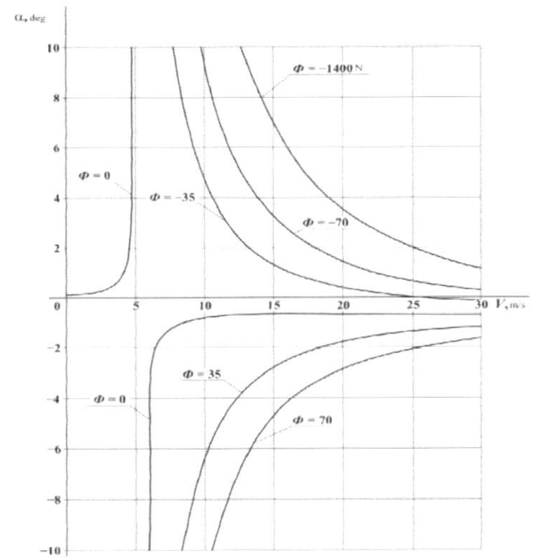

Figure 2. Dependence of the balancing angles of attack from the speed of established horizontal motion

3. Longitudinal stability of the airship

In order to analyse the longitudinal stability of the airship and, as result, possibility to stabilize the airship motion certain equations describing the turbulented moves should be composed. Upon that we will use the Lyapunov theorem of asymptotic stability of established motion. For the particular case described in this article the conditions of asymptotic stability of established longitudinal motion can be aggregated as follows:

$$-\bar{y}_c > \frac{\nu^2}{\mu^2+\mu k_{22}}\left[1-\frac{c_y^\alpha\left(\mu\bar{J}_z+k_{66}\right)}{m_z^{\bar{\omega}_z}\left(\mu+k_{22}\right)}\right]\left[m_z^\alpha\left(2\mu-c_y^{\bar{\omega}_z}\right)+c_y^\alpha\left(m_z^{\bar{\omega}_z}-2\mu\bar{x}_c\right)\right]$$

where

$$\bar{x}_c=\frac{x_c}{l};\ \bar{y}_c=\frac{y_c}{l},\ l=U^{1/3};\ \bar{J}_z=\frac{J_z}{ml^2},\ \mu=\frac{m}{\rho U},\ \nu=\frac{V^*}{2\sqrt{gl}}.$$

4. Conclusions

A mathematical model of the longitudinal steady motion of an airship is designed. The conditions for longitudinal stability of rectilinear horizontal motion are obtained. The values of the speed at which the established straight-line motion of the airship is impossible are found.

References

[1] Khoury G A and Gillett J D 1999 *Airship Technology* (Cambridge: Cambridge University Press)
[2] Khoury G A 1998 Review of unconventional design *Proc. of the 24th Joint Annual Symposium of the Airship Association*, June 26–28, Bedford. 21–42 (1998)
[3] Nayler A. *Airship activity and development world-wide 1998. Proc. of the 24th Joint Annual Symposium of the Airship Association*, pp 1–19.
[4] De Laurier J D 1972 A stability analysis for tethered aerodynamically shaped balloons *J. Aircraft* **9** 646-651
[5] De Laurier J D, Lowe J D, Uffen D R and Amann J H 1986 Research on airship flight dynamics at the university of Toronto *Proc.Conf. on "Airship design and operation – present and future"* (Royal Aeronautical Society)
[6] Fichtl G H 1978 *Problems in simulation of atmospheric boundary layer flows* (AGARD)
[7] Houbolt J C 1973 *Survey on effect of surface wind on aircraft design and operation and recomendations for needed wind research.* NASA CR-2360, **74**
[8] Ramos J, Bueno S, Maeta S, de Paiva E et al. Project AURORA *The 24th Joint Annual Symposium organized by the Airship Association,* 26-28 June 1998, Bedford, England, pp 91-104.
[9] Oswald W.B. 1932 *The Transverse Force Distribution on Ellipsoidal and Nearly Ellipsoidal Bodies Moving in an Arbitrary Potential Flow. Thesis.* (Calif. Institute of Technology)
[10] Bicho E, Moreira A, Carvalheira M and Erlhagen W. 2005 Autonomus *Flight Trajectory Generation via Attractor Dynamics. International Conference on Intelligent Robots and Systems IEEE/RSJ*, pp 615-621
[11] Bestaoui S Y 2012 *Lighter than Air Robots, Intelligent Systems, Control and Automation* (Springer)
[12] Knoblock F D and Troller Th 1935 *Tests on Effect of Sidewind on the Ground-Hanling of Airships* (Daniel Guggenheim Airship Institute).
[13] Ashford R L 1986 Military & civilian applications for tethered aerostats *Conf. on Sounding Rockets, Ballons & related Space Systems, AIAA 7th, Oct 1986, Ocean City Md, Collection of Technical Papers, NewYork NY*, pp 28-34.
[14] Thompson N 1972 Turbulence measurement over the sea by a tethered-balloon technique. *Quart. J. R. Met. Soc.* **98**, 745-762
[15] Rice C L, Korn A O, McPhetres G H and Hunsinger F V 1976 Some unique applications for tethered balloon systems. *AIAA*, **81**

[16] De Laurier J D 1977 *Prediction of Tethered-Aerostat Response to Atmospheric Turbulence. Engineering Notes* (Amsterdam, Netherlands)
[17] Jones S P and Krausman J A 1982 Nonlinear Dynamic Simulation of a Tethered Aerostat *Journal of Aircraft*, **19**, 679-686
[18] Jones S P 1989 *Prediction of aerostat & airship mooring mast loads by nonlinear dynamic simulation. AIAA Lighter-Than-Air Systems Technology Conference, 8th, Jacksonville, Fl, Oct. 5-7, 1989,* Technical Papers, Washington, AIAA
[19] Grumondz V and MatusV. Dynamics of an airship connected to earth with a rope. *The 24th Joint Annual Symposium organized by the Airship Association, 26-28 June 1998, Bedford, England* pp 73-80.
[20] Zasolov R 1985 Aerodynamics model of single-bodied airship *TsAGI* **2268** 31–46 (In Russian)
[21] Arie M 1986 *Airships* (Kiev: Naukova dumka) (In Russian)
[22] Boiko Y. 2001 *Aeronautics* (Moscow. MGUP) (In Russian)
[23] Kirilin A 2013 *Airsips* (Moscow.: MAI-Print) (In Russian)
[24] Grumondz V. and Rudnev V. 1989 Airship established horizontal circulation *22-nd Tsiolkovsky readings* (Moscow: IIET of the the USSR Academy of science) pp 103–107 (In Russian).
[25] Grumondz V and Matus V 1999 Dynamics of the airsip towed by the cable *MAI Vestnik* **6**(2) 60–66 (In Russian).
[26] Grumondz V, Sementhikov N and Yakovlevsky O 2017 *Aeromecanic of the Airship* (Moscow: Nauka) (In Russian)
[27] Loytsansky L 1970 *Liquid and gas mechanics* (Moscow: Nauka) (In Russian)

IOP Conf. Series: Materials Science and Engineering **476** (2019) 012014 doi:10.1088/1757-899X/476/1/012014

The Influence of Solar Radiation on the Parameters of the Polymer Type Test Pattern, Designed to Adjust the Optical Aircraft Equipment

A P Kondratov[1], A A Nikolaev[1], M A Kondratova[2] and I A Savenkova[3]

[1]Moscow Polytechnic University, Moscow 107023, Russia
[2]Moscow State University of Geodesy and Cartography, Moscow 105064, Russia
[3]National Research University Higher School of Economics, Moscow 101000, Russia

E-mail: apkrezerv@mail.ru

Abstract. The flexible compound materials of test objects (type test patterns) to adjust the aviation optical equipment for the Earth remote sensing obtained by the method of ink-jet printing on banner fabric have been studied. The quantitative regularities of changes in the optical and mechanical properties have been determined, which allow predicting the maintenance of the operational characteristics of the type test pattern.

1. Introduction

Adjustment and determination of the spatial resolution of aviation electrooptical equipment is performed by shooting and subsequent recognition of line images applied to objects located in a certain way on the surface of the Earth [1].

Dismountable sets of fabrics made of polymer composite materials [2], which are flexible and resistant to weather factors, are proposed as carriers of line images (type test patterns): strength in dry and wet conditions, thermal and cold resistance. The most important property of the line image carriers is the stability of their optical characteristics. They include printing density or light reflection factor and contrast ratio of adjacent line images forming the type test pattern [3]. To determine the sensitivity limits of aviation electrooptical equipment, it is necessary to manufacture and study type test patterns with low-observable lines characterized by the contrast ratio of about 0.2.

As a result of short-term tests (radiation by mercury lamp with UV light in the laboratory for 100 hours) that have been conducted earlier [4] to assess the light resistance of the type test patterns printed by the ink-jet printer on banner fabric coated with polyvinyl chloride, no significant changes in optical density have been found. Determination of the contrast ratio of adjacent images of light and dark lines and changes in the contrast ratio influenced by UV radiation has not been performed. Therefore, in order to substantiate the reliability of light resistance prediction and ensure the multiple use of the type test patterns made of banner fabric, it is necessary to conduct longer tests with assessing the contrast ratio of the printed images and controlling the mechanical characteristics of the fabrics under conditions similar to operating conditions in areas with maximum solar radiation.

The paper objective is to determine the resistance to solar radiation of low contrast monochrome images of the type test patterns made by ink-jet printing on high-strength banner fabric.

Material: S-PRW334240B banner fabric made in the Republic of Korea from polyester thread grid coated with plastisol of polyvinyl chloride.

Ink for ink-jet printing (solvent ink): BIGINKT-900 BLACK.

The procedure for accelerated testing of the type test pattern corresponds to GOST 28202-89 (IEC 68-2-5-75) Part 2. Tests: Simulated solar radiation at the ground level. The test procedure on option A assumes an 8-hour radiation cycle. Radiation time is 60 days.

The UV source is a lamp of the brand DRT-1700 (SKTB "Xenon") with a capacity of 1700 W that provides a density of ultraviolet light power of 1120 W/m^2. The surface temperature of the polymer coating of the type test pattern is 60°C. It provides a specially designed and manufactured system of temperature control and water cooling of the underlying surface. The method for determining the mechanical characteristics is GOST R 55408-2013. The method of measuring the adhesive strength of the ink film [6].

Determination of optical density and reflection factor of the built-in light source D65 at the observation angle of 2 degrees has been performed using X-Rite Pantone e-Xact spectrophotometer with Gretag Macbeth Key Wizard V2.5 software. As a sample of white color, banner fabric without ink film has been used. Method for determining the contrast ratio by color coordinates [7]. Method for determining the contrast ratio by reflection factor [8].

2. Results and considerations

Type test pattern to adjust and determine electrooptical systems operating in the visible range of electromagnetic waves is repeating groups of light gray lines against a dark background of width 0.5 ÷ 2.0 m. The lines are printed on banner fabric with a mass of square meter 400 g. The thickness of the prints is 0.3 ± 0.04 mm. Conditional density of the compound material of the fabric is 0.132 g/cm^3.

The optical density of the background and light lines of the type test pattern samples before UV radiation is given in the table. The type test pattern sample taken to test the light resistance is characterized by low contrast ratio [9]. The ink film adhesion on the areas of dark background and light lines exceeds the interlaminar strength of banner fabric (table) and is more than 120N/m^2. Large size of the parts of the dismountable type test pattern [2] cause large linear tensile stresses during their laying and aligning on the ground and stress concentration in the places where the fabrics are hooked for uneven stony ground or woody vegetation remainings. For this reason, after the exposure of the type test pattern to the UV radiation, the strength and tensile elongation have been determined.

If the time of the exposure of the type test pattern to the UV radiation increases, the change in the optical characteristics of the background and lines printed with jet inks on banner fabric is most intense during the initial period of time up to 2 days. The optical density of the light line and the optical density of the background are reduced by 14%. This change reduces the contrast ratio of the type test pattern and brings its image to the sensitivity limit of aviation electrooptical systems.

When calculating the contrast ratio ($\boldsymbol{K_1}$) by the formula (1), it changes by 11 % for 2 days

$$\boldsymbol{K_1} = \frac{\rho_h - \rho_f}{\rho_h + \rho_f}. \quad (1)$$

When calculating the contrast ratio ($\boldsymbol{K_2}$) by the formula (2), it changes by 9.7% for 2 days

$$\boldsymbol{K_2} = \frac{\rho_h - \rho_f}{\rho_h} \quad (2)$$

Where $\rho_h$is the reflection factor of the standard source light of the D65 spectrophotometer by the line, ρ_f is the reflection factor of the standard source light of the D65 spectrophotometer by the background

The change in the optical density of the type test pattern on the first day of radiation seems to be based on the removal of volatile organic substances from the ink film and the carbonization of components prone to damage by UV light. Further reduction of the optical density and increase in the factor of light reflection from the lines and background for 60 days is steady and is described by the logarithmic function $\rho = kln(t) + b$ with a probability R^2= 0,93 ÷ 0,94. (Fig.1)

Extrapolating the linear dependence of the reflection factor on time (Fig. 2) and the time dependence of the contrast ratio of the type test pattern (Fig. 3) for the period of solar radiation from

60 to 120 days, it is reasonable to assume that the contrast ratio of the type test pattern does not drop below the sensitivity limit of the electrooptical equipment of aircrafts during summertime.

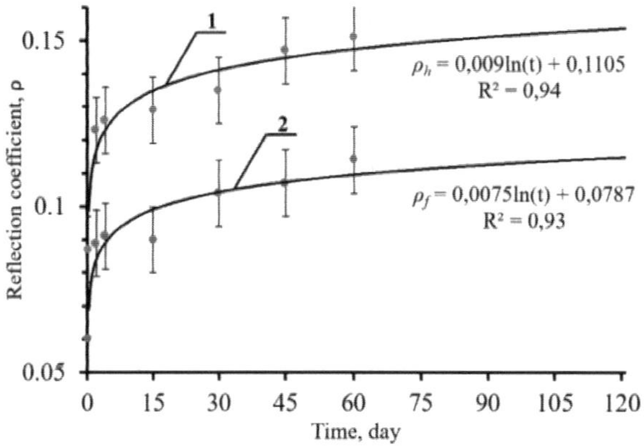

Figure. 1. Dependence of the sunlight reflection factor on the printed lines (1) and dark background (2) of the type test pattern

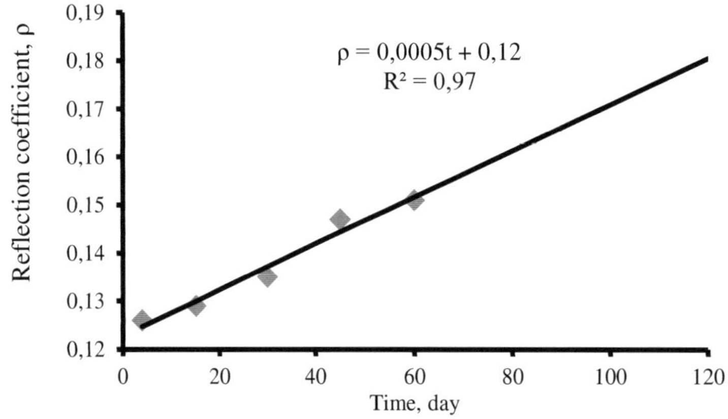

Figure. 2. Linear approximation of the dependence of the sunlight reflection factor on the printed lines (1) of the type test pattern

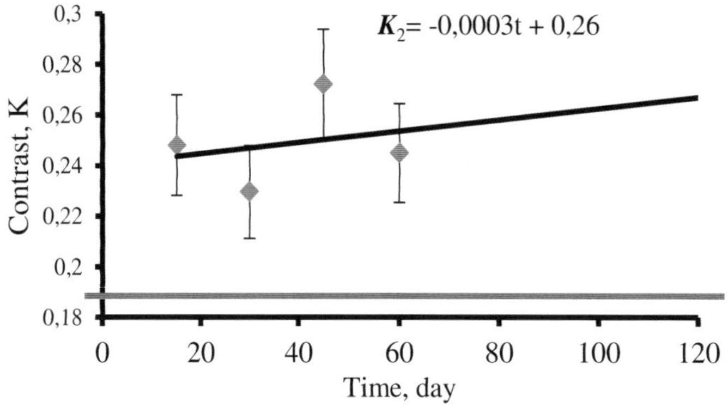

Figure. 3. Linear approximation of changes in the contrast ratio of the type test pattern when radiated with UV light.

The regularities of change in the strength of the type test pattern when irradiated with UV light are different from the regularities of change in the optical properties. During the initial period of solar radiation (up to 15 days), the material strength of the type test pattern decreases slightly (2%). This change in strength is within the limit of errors of the breaking stress measurement by the standard method. In the next 15 days, i.e. after 30 days of radiation, the strength decreases by 20%, and after 45 days of radiation it decreases by 40%. Upon the completion of the test cycle, when the type test pattern is exposed to radiation for 60 days, residual strength of the parts is 24 mPa.

The loss or decrease in the adhesion of the ink film to the polymer coating is a certain danger to the stability of the photometric characteristics of the dismountable type test pattern on the printed banner fabric. When modules of the type test pattern are rolled and stored for a long time, ink that partially destroyed by solar radiation can glue to the back of the banner fabric. To predict and prevent such damage to modules of the type test patterns, the film ink adhesion strength on the printed coating is determined by the method of pull-off [6]. It is impossible to assess the adhesion of the ink film to the banner fabric before the exposure to solar radiation, since it exceeds the interlaminar strength of the fabric. An attempt to separate the ink film results in its delamination, so the table shows the value of the ink adhesion more than 1,2 MPa. When the type test pattern is radiated for 15 days, the ink film can be separated from the banner fabric with normal stress of more than 0,35 MPa. This value does not change significantly throughout the time of accelerated climatic tests and is more than 3 times higher than the standard interlaminar strength of printed products and laminated packaging materials [9].

Table 1. The value of the ink adhesion.

Time of Radiation with UV light, day	Reflection factor ($\boldsymbol{\rho}_{omp}$)		Contrast ratio ($\boldsymbol{K}_2$)	Tensile strength, MPa		Ink film adhesion, MPa	
	line	background		line	background	line	Background
0	0.087	0.06	0.31	40	43	>1.2	>1.2
2	0.123	0.089	0.28	40	43	>1.2	>1.2
15	0.129	0.097	0.25	39	42	53	0.57
30	0.135	0.104	0.23	32	33	0.35	0.50
45	0.147	0.107	0.27	29	31	0.30	0.45
60	0.151	0.114	0.25	24	25	0.25	0.35

3. Conclusion

The analysis of the accelerated testing results for the photochemical stability of a polymer type test pattern sample, designed to determine the spatial resolution of aviation and/or space electrooptical equipment, has established quantitative regularities of changes in optical and mechanical properties, allowing to predict the maintenance of the operational characteristics of the type test pattern at an adequate level.

References

[1] Goryl F and Burini A 2009 *Calibration Test Sites and Cal/Val Portal Renovation. IVOS 21* (University of Lethbridge, Canada)

[2] Altukhov E V, Bablyuk E B, Veselov U G, Vladychenko O V and Moroz V A 2017 Pat. 175973, RF, *Dismountable test object, i.e. type test pattern to determine the parameters of*

electrooptical systems IPC G01M 11/02, G03B 43/00, H04N 17/00, pubd. 25.12.2017, Report No.36 dated 17.04.2017.

[3] Ermakova I N, Kondratov A P and Nagornova I V 2015 Transparent polyolefin-based laminated materials with variable colouring, *International Conference on Oil and Gas Engineering* pp 101-103

[4] Bablyuk E B, Berlad Yu M, Letyago A G and Kondratov A P 2017 Materials for test objects to configure aviation optoelectronic earth remote sensing systems *MATEC Web of Conferences* **99** 01001

[5] Zhuravleva G N, Nagornova I V, Kondratov A P, Bablyuk E B and Varepo L G 2017 Control of the operational properties of polymer materials with applied type test patterns *Methods and technology of petrochemical and oil and gas production materials 7th International Scientific and Technical Conference* pp 185

[6] Kondratov A P, Bozhko N N, Dryga M A and Bablyuk E B 2010 Pat. 2390004, RF, *Method for assessing the connection strength of screen inks and coatings with printing materials* IPC G01N19 / 04, pubd. 20.05.2010, Report No.14 dated 20.05.10

[7] Kondratov A P, Varepo I G, Nagornova I V and Ermakova I N 2015 Transparent layered materials based on variable color polyolefins *Procedia Engineering*, **113** 423–428

[8] Aisenberg U B (Eds.) 2006 *Reference book on lighting engineering* (Moscow: Znak)

[9] Travnikova N P 1985 *Efficiency of visual search* (Moscow: Mashinostroeniye)

IOP Conf. Series: Materials Science and Engineering **476** (2019) 012015 doi:10.1088/1757-899X/476/1/012015

Methods for Calculating Competitive Indicators of High-tech Products, Considering the Factors of Product Sales in the Markets in Aerospace Industry

Y N Koptev[1] and A A Chursin[2,3]

[1] State Corporation «Rostec», Moscow 119048, Russia
[2] JSC Russian Space Systems ,Moscow 111250, Russia
[3] Peoples' Friendship University of Russia, Moscow 117198, Russia

E-mail: achursin2008@yandex.ru

Abstract. The article suggests a modelling tool for assessing competitiveness in hi-tech entreprises including aerospace industry. According to the primer approach in the theory of management of high-tech industries competitiveness, it is necessary to use the strategy for managing various factors, arising in the global and domestic markets according to the knowledge of their interdependencies. But the statistical approaches have the limited forecasting force when the sales of a certain type of hi-tech product are unique or no experience was obtained before. Nevertheless, the uncertainty can be seriously diminished if we use mathematical modelling based on well-calibrated functions which depend on our hypotheses and theoretical assumption. In the paper a possible principle of theoretical modelling the competitiveness is outlined.

1. Introduction

The competitiveness of high-tech products is significantly influenced by the factors that arise during the sale of these products in the global and domestic markets [1-2]. In fact, the real competitiveness of products becomes obvious during the sale of this product. According to the primer approach in the theory of management of high-tech industries competitiveness, it is necessary to use the strategy for managing various factors, arising in the global and domestic markets.

To take into account the factors that arise in the process of sale, it is necessary to manage the products competitiveness, which in the context of high-tech industries necessitates decision-making under conditions of uncertainty. Uncertainty arises from market processes, the sale of products and competition with other products) [3-5].

There exist some approaches to measure the competitiveness. The first group of methods is connected with the measuring the actual market data, market shares in dynamics [6]. If the series of sales are great by number, the statistical methods are applied, and they may be used for forecasting. But they have the limited forecasting force when the sales are unique (maybe very customized product), when no sales history was recorded before (for example, the unique product is sold by the first time),the product is completely new (as it is often in aerospace industry). So, in this case, the possibility of market measuring and statistics methods is limited. Here exists an uncertainty which could not be overcome completely. Nevertheless, the uncertainty can be seriously diminished if we use mathematical modelling using functions which depend on our hypotheses and based on proper theory. Though hypotheses are chosen totally by a modeler externally, the model itself can show the interaction between the hypothesis and lead to the conjecture based upon strict premises. Below we outline a possible principle of modelling the competitiveness.

2. Methods

The mathematical model of accounting internal and external factors, arising in the process of their sale in the global and domestic markets can be presented in the following form.

A set of products, competing with the considered product in the domestic market is considered:

$$N_i, \quad i = 1, 2, \ldots, n.$$

A set of products, competing with a product in the global market is also considered:

$$N_i, \quad i = 1, 2, \ldots, n.$$

Let the Q indicate the integral index of competitiveness, calculated on the basis of sales in the domestic and global markets. In this case, the competitiveness indicator significantly depends on the price of a product and on the prices of competing products. Let us consider the equation that defines the integral index of competitiveness:

$$Q = \Phi_{in}(C_{in}, C_{in}^1, C_{in}^2, \ldots, C_{in}^n) + \Phi_w(C_w, C_w^1, C_w^2, \ldots, C_w^m), .$$

where Φ_{in} is a functional of competitiveness in the domestic market; Φ_w is a functional of competitiveness in the global market; C_{in} is a cost of products in the domestic market; C_{in}^i is a cost of competing products in the domestic market; C_w is a cost of products in the global market; C_w^j is a cost of competing products in the global market.

Let us consider 3 groups of factors, affecting the competitiveness of products:

- factors, determined by the technical and economic properties of products;
- economic factors (price);
- random factors.

Economic indicators are included in the cost of products in the domestic and global markets.

Technical and economic properties of products affect the integral index of competitiveness. To account the indicators that determine the technical and economic properties of products, let us introduce the parameters for accounting technical and economic properties:

- a_{in} is a parameter of accounting technical and economic properties of products in the domestic market;
- a_w is a parameter of accounting technical and economic properties of competing products in the global market;
- a_{in}^i is a parameter of accounting technical and economic properties of competing products in the domestic market;
- a_w^j is a parameter of accounting technical and economic properties of competing products in the global market.

The greater value of the parameter of accounting technical and economic properties of products corresponds to a greater value of the final index of its competitiveness.

Finally, random internal and external factors influence the final indicator of the competitiveness of high-tech industries. To account such factors, let us introduce a random value of change in the competitiveness under the influence of random factors:

- ρ_{in} is a coefficient of change in the competitiveness of products under the influence of random factors in the domestic market;
- ρ_w is a coefficient of change in the competitiveness of products under the influence of random factors in the global market;
- ρ_{in}^i is a coefficient of change in the competitiveness of competing products under the influence of random factors in the domestic market;

- ρ_w^j is a coefficient of change in the competitiveness of competing products under the influence of random factors in the global market.

The coefficient of product competitiveness change under the influence of random internal and external factors influences the final value of the integrated indicator of the competitiveness of high-tech industries as follows: the increase in ρ_{in} and ρ_w indicators raises.

Thus, after considering the factors, affecting the competitiveness of products, the integral indicator of competitiveness will be determined by the following expression:

$$Q = \Phi_{in}[q_{in}(C_{in}, a_{in}, \rho_{in}), q_{in}^1(C_{in}^1, a_{in}^1, \rho_{in}^1), \dots, q_{in}^n(C_{in}^n, a_{in}^n, \rho_{in}^n)] +$$
$$+\Phi_w[q_w(C_w, a_w, \rho_w), q_w^1(C_w^1, a_w^1, \rho_w^1), \dots, q_w^n(C_w^m, a_w^m, \rho_w^m)],$$

where $q_{in}(C_{in}, a_{in}, \rho_{in})$ is a function of private competitiveness of products in the domestic market; $q_w(C_w, a_w, \rho_w)$ is a function of private competitiveness of products in the global market; $q_{in}^i(C_{in}^i, a_{in}^i, \rho_{in}^i)$ is a function of private competitiveness of competing products in the domestic market; $q_w^j(C_w^j, a_w^j, \rho_w^j)$ is a function of private competitiveness of competing products in the global market;

The influence of the coefficient of change in product competitiveness under the influence of random internal and external factors on the value of the functions of private competitiveness is as follows: an increase in the ρ indicator raises the value of the function of private competitiveness. Analogically, an increase in the a parameter of accounting technical and economic properties of products leads to an increase in the value of the function of private competitiveness of products.

Let us now consider the impact of the values of the functions of private competitiveness on the final value of the integral indicator of competitiveness. To consider the quantitative effect of this value, it is necessary to consider the corresponding derivatives.

Derivative

$$\frac{\partial Q}{\partial q_{in}} = \frac{\partial \Phi_{in}[q_{in}(C_{in}, a_{in}, \rho_{in}), q_{in}^1(C_{in}^1, a_{in}^1, \rho_{in}^1), \dots, q_{in}^n(C_{in}^n, a_{in}^n, \rho_{in}^n)]}{\partial q_{in}} > 0.$$

It means that the growth of the value of the function of private competitiveness of high-tech products in the domestic market leads to the increase of the value of the integral indicator of competitiveness.

Derivative

$$\frac{\partial Q}{\partial q_w} = \frac{\partial \Phi_w[q_w(C_w, a_w, \rho_w), q_w^1(C_w^1, a_w^1, \rho_w^1), \dots, q_w^m(C_w^m, a_w^m, \rho_w^m)]}{\partial q_w} > 0.$$

It means that the growth of the value of the function of private competitiveness of high-tech products in the global market leads to the increase of the value of the integral indicator of competitiveness.

Derivative

$$\frac{\partial Q}{\partial q_{in}^i} = \frac{\partial \Phi_{in}[q_{in}(C_{in}, a_{in}, \rho_{in}), q_{in}^1(C_{in}^1, a_{in}^1, \rho_{in}^1), \dots, q_{in}^n(C_{in}^n, a_{in}^n, \rho_{in}^n)]}{\partial q_{in}^i} < 0.$$

It means that the growth of the value of the function of private competitiveness of products in the domestic market leads to the reduction of the value of the integral indicator of competitiveness.

Derivative

$$\frac{\partial Q}{\partial q_w^j} = \frac{\partial \Phi_w[q_w(C_w, a_w, \rho_w), q_w^1(C_w^1, a_w^1, \rho_w^1), \dots, q_w^m(C_w^m, a_w^m, \rho_w^m)]}{\partial q_w^j} < 0.$$

It means that the growth of the value of the private competitiveness function of the competitive product in the global market leads to the reduction of the value of the integral indicator of competitiveness.

3. Conclusion

Methods of quantitative assessment of the competitiveness of products, taking into account a variety of external and internal factors [7-8], is a necessary link in the analysis and planning of economic activity of high-tech industries in modern conditions, formed in the global and domestic markets during the sale of high-tech products. External and internal factors, affecting the competitiveness of products, while their sale in the global and domestic markets, have a fundamental impact on competitiveness indicators in the context of the global crisis and post-crisis phenomena in the global economy, as well as the Russian economy [9].

Specific character of high-tech industry products is that there is relatively small number of the same types of products in the domestic and foreign markets. Moreover, science-intensive industries have a very high barrier of entry into the global and domestic markets. Consequently, the factors that arise in the sale of products in the global and foreign markets can be calculated, using the analysis of the same type of products.

Methods for assessing the competitiveness of products, taking into account the factors that arise in the sale in the global and domestic markets, solve the problem of effective monitoring of the progress of high-tech products competitiveness. This task is necessary for the permanent management of the high-tech industries competitiveness.

Acknowledgments

This paper was financially supported by the Ministry of Education and Science of the Russian Federation on the project No.26.1146.2017/4.6 «Development of mathematical methods to forecast efficiency of using space services in the national economy».

References

[1] Chursin A & Makarov Yu 2015 *Management of Competitiveness. Theory and Practice.* (Springer International Publishing).

[2] Tyulin A, Chursin A and Yudin A 2017 Production capacity optimization in cases of a new business line launching in a company *Espacios* **38** 62

[3] Lemmink J and Kasper H 1994 Competitive Reactions to Product Quality Improvements in Industrial Markets *European Journal of Marketing* **28**(12) 50-68

[4] Easton G 1988 Competition and Marketing Strategy *European Journal of Marketing* **22**(2) 31-49

[5] Javalgi R G, Radulovich Lori P, Pendleton G, Scherer Robert F 2005 Sustainable competitive advantage of internet firms: A strategic framework and implications for global marketers *International Marketing Review* **22**(6) 658-672

[6] Mazzeo Michael J 2002 Competitive outcomes in product-differentiated oligopoly. *The Review of Economics and Statistics* **84**(4) 716–728

[7] Kashirin A, Semenov A, Ostrovskaya A, and Kokuytseva T 2016 The Modern Approach to Competencies Management Based on IT Solutions *JIBC-AD - Journal of Internet Banking and Commerce* **01** 1-12

[8] Kashirin A, Semenov A, Ostrovskaya A, Kokuytseva T and Strenaluk V 2016 The Modern Approach to Competence Management and Unique Technological Competences *Quality – access to success* **17**(154) 105-109

[9] Chursin A, Semenov A and Danilchankau A 2016 Analysis of Innovation Development in the Economy with Exhaustible Resource Sector by First Order Dynamical Systems Application *Nonlinear phenomena in complex systems* **19**(3) 254-270

IOP Conf. Series: Materials Science and Engineering **476** (2019) 012016 doi:10.1088/1757-899X/476/1/012016

Industrial Informatics: Towards the Researches and Modelling of Structural Materials' Properties

E E Kovshov[*], **A A Kosach and V S Popov**

JSC «NIKIMT-Atomstroy», Moscow, Russian Federation

[*]E-mail: eugene.kovshov@gmail.com

Abstract. Considers the possibility of applying the principles of Industrial Informatics to optimize the study and modelling of various structural materials and products on the example of intelligent software and hardware platform based on mathematical apparatus and modules. The ways of practical implementation of mathematical and bio-inspired algorithms and models for the developed software applications in the corporate information system of industrial enterprise paradigm-based Industry 4.0 are proposed.

1. Introduction

Industrial Informatics as an independent scientific and engineering direction appeared at the junction and as a result of the rapid development of science, technology and computer science. The term "industrial" refers to the creation of real industrial software applications, and the term "informatics" refers to the infrastructure that provides development, implementation and maintenance throughout the life cycle of existence of it. In addition, Informatics as a science offers tools and methods for analysis, processing, transformation and transfer of information. Industrial Informatics focuses primarily on the automation based on knowledge as a means of re-engineering and modification of production processes in industry.

Industrial Informatics is not limited to metal working, for example, there are such branches of knowledge as computer control systems, robotics, intelligent video surveillance and image processing systems, as well as data collection of different nature and processing of multidimensional signals, where mathematical methods and tools of Informatics are widely used.

Industrial Informatics has a set of techniques and practices that use information analysis, processing and dissemination to achieve more meaningful results in terms of efficiency, reliability and/or safety in an industrial environment. The field of industrial Informatics has become one of the key areas for intelligent control and digital production technologies.

Information Technology (IT) design tools for different industries vary widely depending on the nature of the industry and, accordingly, the business processes being implemented.

For example, the use of IT in manufacturing industry includes modeling business processes, planning and production control, distributed information system inventory management and knowledge management for applied research.

Industrial Informatics – "building", built on the foundation of "classical" Informatics using the theory of algorithms, fuzzy logic, artificial neural networks, artificial immune systems, evolutionary and genetic algorithms, industrial design of information systems, as well as accumulated considerable practical experience in the field of theoretical and applied Informatics. Industrial design processes are defined as a set of logically related tasks performed to achieve a specific result, with the industrial production process defined based on the customer's requirements and business interests.

2. Digital intelligent platform

One of the main trends in the development of application software in the structure of digital production was the transition from monolithic construction of information systems to modular, which greatly facilitates the scaling of the software.

The last stage in the development of modular programming was the service-oriented architecture (SOA) of information systems [1]. SOA is a modular approach to software development based on the use of distributed, loosely coupled standardized components (services). Services are compatible, flexible and reusable. Accordingly, all the functionality of the system is implemented in components that are built in such a way that they are nearly unrelated to each other. The introduction of information systems based on SOA becomes preferable with the growing scale of data and calculations.

This intelligent software platform for data collection and processing with non-destructive testing is based on service-oriented architecture of software solutions and is one of the information subsystems in the structure of integrated digital production.

Developed and tested (as part of the SOA platform) intelligent software platform for data collection and processing of the automated information-measuring system of leakage testing [2]. The services perform various application tasks (e.g. image recognition, strength test calculation) and are combined with each other only to solve the problems identified by their application shown on the model (figure 1).

It should be noted that business processes serve as a mechanism for ensuring the interaction of services with each other. A business process is a set of interrelated tasks and it controls the flow of events, calls and coordinates services, and creates a context for their interaction. A business process is an abstract mechanism that does not depend on how services are implemented and what logic they carry. The main task of the business process is the organization of services for its effective implementation.

The orchestration module is designed to select a business process that interacts with different services. It defines a mechanism for service interaction based on business process logic and represents a sequence of actions performed by services. The enterprise bus (in this case, ESB) provides an environment for interacting with the service.

One of the most efficient ways to implement a service-oriented approach is to use web-services, where the service-oriented model covers all levels of data management in a software platform based on SOA. It also has no distribution to suppliers and consumers, as all the necessary information about the services is in the register. For more complex and in-depth data analysis, services located in a public, private, or hybrid cloud can be used.

Open ESB corporate service bus was used for implementation of information-measuring control system [3]. Services are implemented in the form of web-services, the functional part of which is written in Java. The system elements exchange messages based on the SOAP Protocol [4]. To store data and calculation results an open-source relational database management system is used. The use of these technologies allows you to analyze and filter data, including in the main Microsoft Office software (for example, Microsoft Excel, Microsoft Access).

For the exchange of information between parts of the information-measuring system applied OPC servers (OPC UA, OPC DA) and the communication protocols Modbus RTU and Modbus TCP. OPC is a technology based on the use of a single device management interface [5], this is especially important for automating remote control systems. Often, developers of automated systems have to include a lot of ready-made drivers in the control system or use the tools to develop original communication protocols with non-standard devices of the lower level. And when they replace a device with a similar one from another manufacturer, they need to significantly change the software to control that device. A single OPC interface allows different software modules produced by different companies to interact with each other.

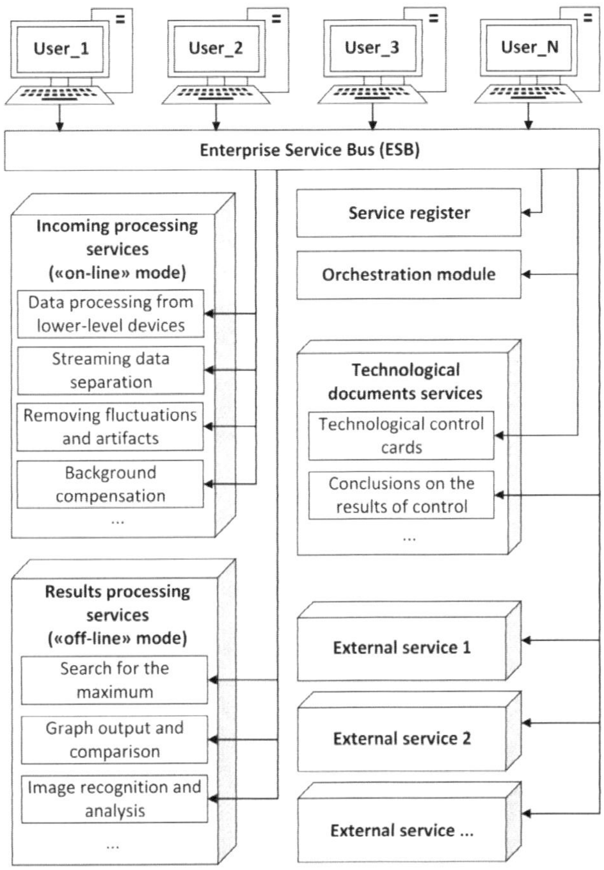

Figure 1. The layout of the software platform based on SOA.

3. Mathematical apparatus

This software platform from our point of view is applicable for the selection of structural materials with certain properties in the design of products and to reduce the number of options to be modeled and calculated. Modeling and calculation of products requires significant time and computer resources, and the preliminary selection of the most preferred options from the set reduces these costs. It is also possible to significantly improve certain characteristics of the product, making it, for example, from composite materials or layers of different materials, and the use of algorithms for optimization allows you to select solutions from a very wide range of possible.

At the moment, these tasks are performed mainly using genetic algorithms [6]. Genetic algorithm is an adaptive method of finding solutions. It uses the mechanisms of genetic inheritance, natural selection, biological terminology and basic concepts of linear algebra. The main idea of the genetic algorithm is the organization of natural selection among a variety of solutions to find the extremum of the objective function [7].

Besides this method, you can solve the problem of selecting the required structural material using a neural network or cluster analysis. These methods also allow you to select the most appropriate option from the set with a large number of specified conditions.

Depending on the task, it is advisable to optimize the selection of material on the basis of a neural network module with training with a teacher. The neural network consists of a multilayer perceptron [8], which includes: n layers of k neurons transmitting information through channels with weights (w_{ij}- the weight of the connection between neurons i and j), d inputs and one output, which gives the result (figure 2).

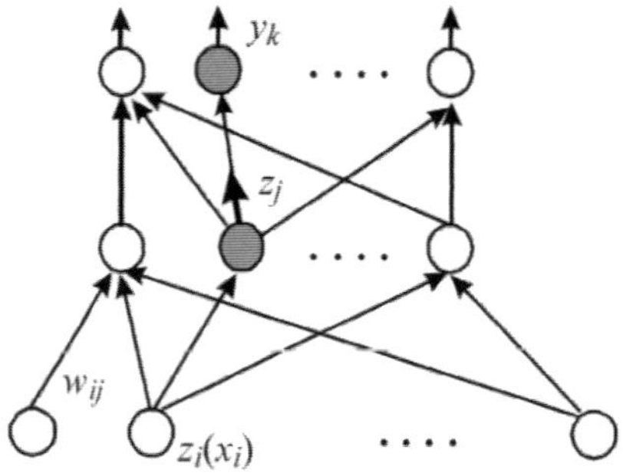

Figure 2. A fragment of the artificial neural network.

When information is received at the inputs, it is processed by the first layer of neutrons, each of which calculates the weighted sum of its inputs:

$$a_i = \sum_n w_{ij} z_i \qquad (1)$$

where z_i is the output of neuron i and the input of neuron j, then the output of neuron j is the conversion of a_j by the threshold function of this neuron:

$$z_j = g(a_j) \qquad (2)$$

The training of an artificial neural network is to determine the weights for each bonds. When training with the teacher, a sample of input vectors $\{x_n\}$ is used, for which the desired result of the neural network $t \in \{t_n\}$ is known (for example, the results of modeling products from certain materials), which allows you to effectively adjust the network weights.

Also, for the selection of materials with the necessary properties, SVM-classifier (SVM - Support Vector Machines) is used [9], which implements training on precedents (analog of training with a teacher). This classifier is advisable to use for the distribution of a large number of solutions (in this case, with respect to materials for the manufacture of certain products) for classes that meet a set of specified characteristics.

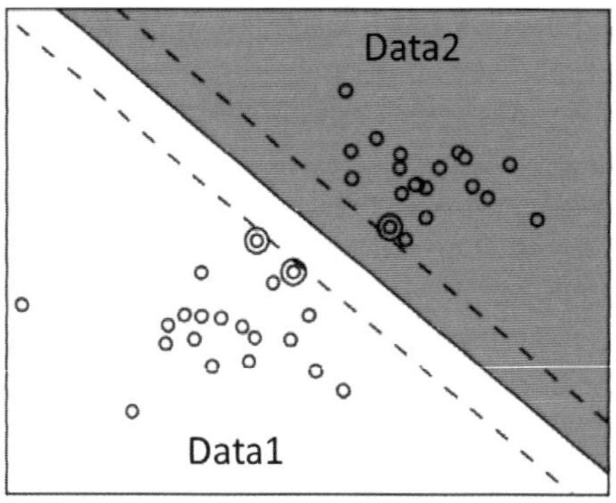

Figure 3. The result of separating two-dimensional data by SVM classifier.

Consider the two-dimensional option. Each characteristic of classified objects is represented as a q-dimensional vector of numerical values of characteristics $z_i = (z_{i1}, z_{i2}, ..., z_{iq})$, where z_{iq} is the numerical value of characteristic q for object i and each of these vectors must belong to only one of the two classes. Then the search for the separating hyperplane with the maximum gap in this space is carried out (figure 3).

There is a data set $\{(z_1, y_1), ..., (z_n, y_n)\}$, where each object z_i of the set $\boldsymbol{Z}$ corresponds to the number y_i of the set $\boldsymbol{Y}$, which takes the value "-1" or "+1", depending on the class to which the object z_i belongs. Required to create a classifier F that associates a random object from the set $\boldsymbol{Z}$ with a class from the set $\boldsymbol{Y}$ using the kernel function k. As a kernel function that allows you to separate objects of different classes, different functions are used (for example, linear or polynomials).

After training, the classification function is determined:

$$f(z)=\sum_{i=1}^{N}\alpha_i \cdot y_i \cdot \kappa(z_i, z)+b \tag{3}$$

where k is a function of the classifier kernel; b is a parameter that determines the shortest distance from the beginning to the hyperplane; α_i - Lagrange multiplier; y_i is a classification solution.

When developing an SVM classifier, the source data (training set) is randomly divided into training and test samples. The training sample is used for training the classifier. The test sample (some quantity of objects of the training set) is not used for training the classifier, and is used to verify the accuracy of the classifier. The result of learning the SVM algorithm is a classifying function.

The decision of classification, which associates the object with the class with a "-1" or "+1", is determined by the formula:

$$F(z)=sign(f(z))=sign\left(\sum_{i=1}^{N}\alpha_i \cdot y_i \cdot \kappa(z_i, z)+b\right) \tag{4}$$

Denote for a training set $\boldsymbol{U}$ containing a set of $\boldsymbol{Z}$ objects the classification rule A(z). Then the following values the number of outcomes of the correct classification can be determined:

- number of correctly defined objects of class "+1":

$$TP(A, Z^*)=\sum_{z=Z^*}(A(z)=+1 \wedge y^*(z)=+1 \tag{5}$$

- number of correctly defined objects of class "-1":

$$TN(A, Z^*)=\sum_{z=Z^*}(A(z)=-1 \wedge y^*(z)=-1 \tag{6}$$

And the values that determine the number of incorrect classification outcomes [9]:

- number of incorrectly defined objects of class "+1":

$$FP(A, Z^*)=\sum_{z=Z^*}(A(z)=+1 \wedge y^*(z) \tag{7}$$

- number of incorrectly defined objects of class "-1":

$$FN(A, Z^*)=\sum_{z=Z^*}(A(z)=-1 \wedge y^*(z) \tag{8}$$

Relative performance indicators of the classifier can also be formulated for further calculation of the classification quality:

- sensitivity (completeness) is the ratio of correctly defined objects of class "+1" to objects really belonging to class "+1":

$$R(A, Z^*)=\frac{TP(A, Z^*)}{TP(A, Z^*)+FN(A, Z^*)} \tag{9}$$

- accuracy is the ratio of correctly defined objects of class "+1" to objects assigned to class "+1" in the classification process:

$$P(A, Z^*)=\frac{TP(A, Z^*)}{TP(A, Z^*)+FP(A, Z^*)} \tag{10}$$

To assess the quality of the classification, taking into account the accuracy and completeness, it is advisable to use combined metrics. For example, the F_β-measure:

$$F_\beta(A,Z^*)=\frac{(\beta^2+1)P(A,Z^*)\cdot R(A,Z^*)}{\beta^2 P(A,Z^*)+R(A,Z^*)} \tag{11}$$

where β is the coefficient reflecting the expected value of accuracy relative to completeness.

Or balanced metric (F_1):

$$F_1(A,Z^*)=\frac{2P(A,Z^*)\cdot R(A,Z^*)}{P(A,Z^*)+R(A,Z^*)} \tag{12}$$

The balanced metric is a special case of the F_β-measure (β=1) and is equal to the mean harmonic between the accuracy and completeness indices.

The value of F_1 tends to 1 if the accuracy and completeness are close to 1, which shows good quality of the classification. Thus, knowing these indicators it is possible to establish in the course of training whether the given classifier for a certain task is correctly picked up and whether in sufficient volume training was carried out.

4. Conclusion

The current stage of development of society is characterized by a significant impact on it of information and telecommunication technologies, which are increasingly and deeply seep into all spheres of human activity without exception, ensure the dissemination of information in society and, at the same time, forming a global information space. An integral and important part of these processes is the penetration of modern information and communication technologies in the field of industrial production and digital materials science as well.

Significant simplification and cheapening of the industrial production brings widespread use of information technology in combination with various means of mathematical analysis and adaptive methods of finding solutions, including in the modeling and study of the structural materials' properties.

References

[1] Vasiliev Y 2007 *SOA and WS-BPEL* (Packt Publishing Ltd.32)
[2] Kosach A and Kovshov E 2017 Development of a computer based problem-oriented automated remote leakage detection control system *ITM Web of Conference* **10** 01005
[3] *Sun GlassFish Enterprise Service Bus* http://download.oracle.com/docs/cd/E19182-01/index.html
[4] Kosach A and Kovshov E 2017 Automated remote computer monitoring of environmentally hazardous products *The 14th International Conference «Application of Contemporary Non-Destructive Testing in Engineering»* (Slovenia: University of Ljubljana) pp 95-102
[5] Mahnke W, Leitner S and Damm M 2009 *OPC Unified Architecture* (Springer)
[6] Leung M and Nevill G 1993 Genetic Algorithms for Preliminary 2-D Structural Design *AIAA Journal* **94** 2287-2291
[7] Punch W, Averill R and Goodman E 1995 Design using Genetic Algorithms - Some Results for Composite Material Structures *IEEE Expert – Intelligent Systems and their Applications* pp 42-49
[8] Kandel E, Schwartz J and Jessell T 2000 *Principles of neural science* (Appleton & Lange)
[9] Demidova L, Sokolova Yu and Nikulchev E 2015 Use Of Fuzzy Clustering Algorithms' Ensemble For SVM Classifier Development *International Review on Modelling and Simulations* **8** 446-457

IOP Conf. Series: Materials Science and Engineering **476** (2019) 012017 doi:10.1088/1757-899X/476/1/012017

Algebraic Formalization of System Design Based on a Purposeful Approach

V Kulabukhov

Moscow research and production complex "Avionika" named after O.V. Uspenskiy, Moscow, 127055, Russia

E-mail: kws0704@mail.ru

Abstract. The article shows that the modern purposeful approach to the design of systems, which has a clear physical since and logical substantiation, allows to obtain a convenient mathematical description using modern algebra. The purposeful approach not only ensures the implementation of the system goals, but also allows you to formally describe the rules of optimization of the structure at the content level. The methods of decomposition and synthesis of the structure used in this approach to the design of systems are quite simply formalized and justified.

1. Introduction

The method of modern algebra allows obtaining new results in solving numerous problems of system theory [1-10]. Previously, algebraic formalization was used to solve the problems of observation, regulation, control of systems [1, 5, 7-10]. System design problems are difficult to formalize. They are usually solved at the descriptive level. All designed systems belong to the class of target systems, that is, there is always a certain purpose for the design, and the formulation of this purpose is usually strict and definite. Modern resources, on which basis systems are created, have reached a high level of development, allowing the use of an appropriate target approach (top-to-down approach) to the design of systems, that is, allowing to proceed from the goals of the system, and not from the possibilities of resources. The approach ensures the implementation of the system goals, allows justifying logically and formalizing the rules of system design. The essence of the approach is described in [11] and briefly explained below.

Within the framework of the approach, all the basic concepts used in the design of any system are strictly defined and logically linked: concept, technical appearance, technical requirements, system structure, system architecture, resources, efficiency criteria, design principles, restrictions. The main task in the design is to optimize the structure, based on the purpose (goals) of the system. The purposeful approach allows simplifying and determining the process of solving this problem by its natural decomposition.

The structure comprehensively characterizes the method and form of organization of parts of the system as a whole, that is, such an organization of resources that form the system, through which the implementation of system goals is achieved. Resources are understood as a material, energy, information or other form of ensuring the creation, existence and functioning of the system in accordance with its intended purpose. The existence and operation of any system is linked to the unavoidable consumption of resources. The system itself does not reimburse resources and they can be replenished only from the outside. The concept of "resources" is based on the principle of inevitability and irreversibility of their consumption.

Analysis of the entire set of resources involved in the creation of the system shows the possibility of their division into independent, non-overlapping classes. This division is inherent in the very nature of resources and is observed in practice. The presence of a finite set of independent resource classes determines the corresponding number of independent, orthogonal points of view on the system and its structure. Each class of resources strictly corresponds to its own, derived architecture, as an orthogonal projection of the system structure on this class of resources.

The structure in this approach is fully characterized by a strictly defined number of architectures, describing it from various independent, orthogonal points of view. Architecture is some orthogonal "slice", "layer" or "stratum" of the structure. With a targeted approach, the architectures are designed according to the "top-to-down" principle: from the functional architecture reflecting the goals and purpose of the system to the architecture (topology) of the computer network and further to the physical (hardware) architecture and the architecture of the energy supply. A subset of the design choices for the respective classes of resources are evaluated according to the criteria of local selection. Generation of variants of architectures is carried out taking into account the axiomatic principles of design assigned in advance (for example, modularity, openness, etc.) and resource constraints. The combination of the best architecture choices selected in the each class of resources, forms the structure of the system.

The structure is checked for compliance with global selection criteria (e.g. weight, dimensions, energy consumption, etc.). If the criteria are met, the system design process is considered completed. If not satisfied, either the resource constraints are relaxed or the terms of reference (technical appearance and related goals and criteria) are adjusted. The process is iteratively repeated until the required goals are achieved.

It is obvious that the procedure for selecting the best variant of the structure on a set of criteria is the procedure for optimizing the structure. Thus, the purposeful approach not only ensures the implementation of the system goals, but also allows you to logically justify and formally describe the rules of optimization of the structure at the content level. The methods of decomposition and synthesis of the structure used in this approach to the design of systems are quite simply formalized and justified in the language of modern algebra.

The apparatus of modern algebra even in solving traditional problems of system theory allows to obtain new results. Previously, algebraic formalization was used to solve the problems of observation, regulation, control of systems [12-15]. System design problems are difficult to formalize. They are usually solved at the descriptive level. The article shows that the modern purposeful approach to the design of systems, which has a clear meaning and logical substantiation, allows to obtain a convenient mathematical description using modern algebra.

2. Description of the research

In the most general form, any system Σ can be represented by a triple

$$\Sigma = (X, s, Y),$$

where s - is the structure of the system that images the set of inputs X to the set of outputs Y, i.e.

$$s : X \to Y$$

At the design stage, the system is formalized in the form of a triple

$$\Sigma = (R, s, C),$$

in which as a set of inputs R is considered a set of resources on the basis of which the system is expected to implement, and as a set of outputs C is considered a set of goals and related functions that the system must have in accordance with the requirements of technical specification to develop it. Then the structure of s is an image of $s: R \to C$.

Along with the representation of the system in the form of a structure-imaging the set of resources R onto the set of goals C, the system Σ can be defined as a purposeful set ω_s, which is a binary relationship between the sets of resources R and goals C (see Fig. 1), that is $\omega_s \subset \boldsymbol{R} \times \boldsymbol{C}$ and

$\omega_s = R \times C = \{(r,c) \in \omega_s: r \mapsto s(r)=c\}$, and $R \subset \boldsymbol{R}$, $C \subset \boldsymbol{C}$.

Let there exist a partition $R/\sim\ = R/\omega_s$ of the set R of resources into equivalence classes with respect to ω_s. Then (see [16]), as shown by the commutative diagram in Fig. 2, the structure of $s{:}R \to C$ admits factorization $s = \bar{s} \cdot p$, where p - is surjective, $\bar{s}$ - is injective.

Physically, the factorization of the structure s means stratification of system $\Sigma=(R,s,C)$ into independent (non - overlapping) levels of description-strata corresponding to the independent classes of resources identified during factorization and the corresponding classes of targets.

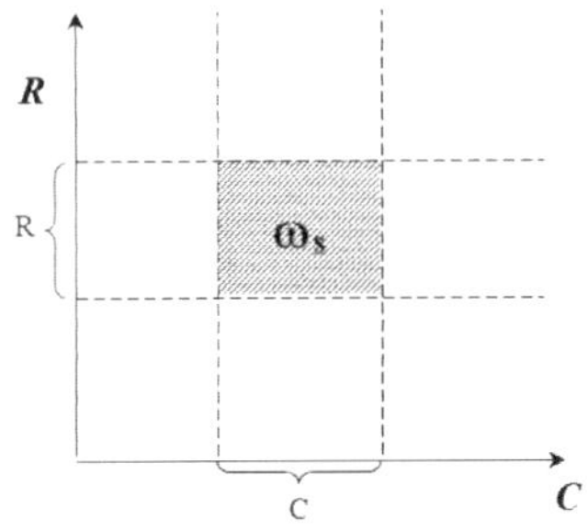

Figure 1. The system, as a binary relation.

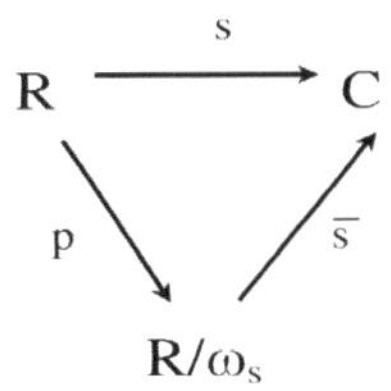

Figure 2. Factorization of system structure.

Definition. Stratification of system $\Sigma=(R,s,C)$ is the set of R/ω_s classes of equivalence of its resources with respect to ω_s.

Statement. Stratification R/ω_s induces in the structure s of system Σ functionally-oriented "non-overlapping" architectures $a \in \bar{s}$.

Indeed, injectivity $\bar{s}$ means that each stratum resources $R_a \subset R/\omega_s$ through architecture $a \in \bar{s}$ is imaged onto the certain subset of purposes $C_a \subset C$ so that

$$\bigcup_a C_a = C,\ \bigcap_a C_a = \varnothing,$$

that is, architecture a: $R_a \to C_a$ is the image of some class of resources onto relevant class of purposes.

Theorem (on factorization of the system structure). Factorization (decomposition) of structure $s = \bar{s}p$ into the composition of surjective p and injective $\bar{s}$ imaging allows to implement the full set of purposes C of system Σ for independent designed architectures $a \in \bar{s}$.

Proving. Truly:

a) for identified stratification of resources R/ω_s image $\bar{s}$ is the only one;

b) if s is surjective (and it was originally due to the requirements of the technical specification, since the structure needs to cover all the set of purposes specified with the task for the development of the system), then $\bar{s}$ is surjective as well, which proves the statement of the theorem.

The following consequence follows from the theorem.

Consequence. The image (structure) $s = \bar{s}p$ is bijective.

Truly:

a) from the limitedness of set R of involved resources and surjectivity of image p onto itself follows the bijectivity of p;

b) from the injectivity and surjectivity of image $\overline{s}$ follows the bijectivity of $\overline{s}$;

c) the composition $s = \overline{s}p$ of the bijective image $p: R \rightarrow R/\omega_s$ and $\overline{s}: R/\omega_s \rightarrow C$ is bijective, and $s^{-1} = (\overline{s}p)^{-1} = p^{-1}\overline{s}^{-1}$. The existence of an inverse image of s^{-1} follows from the bijectivity of s.

Thus, the system can be designed following the principle "down-to-top" from resources to goals, and Vice versa, following the principle "top – to-down" - from goals to resources. That is, from a formal mathematical point of view, both approaches are absolutely equivalent and legitimate. At the same time, in modern conditions, according to a number of additional criteria, a purposeful approach – from goals to resources – is more preferable, since it allows to significantly simplify and to vastly formalize the creative stage of generating options for the structures of the designed system. At the same time, such an approach ensures the implementation of the entire set of goals prescribed by the purpose of the system, and the set of resources involved in the implementation of the goals of the system are considered as limitations, which in modern conditions are significantly softer than before, and for some classes of resources are sometimes completely removed.

The considered consequence allows us to justify the possibility of a purposeful approach to system design:

- its acceptability is due to the bijectivity of structure s and the existence of s^{-1};
- its expediency is due to a simpler procedure of selecting "functional" structure s^{-1}, which is inverse to structure s, compared with the procedure for designing structure s (usually based on combinatorial methods);
- the bijectivity of structure s does not mean its uniqueness, but when a particular structure is fixed, the reverse structure s^{-1} will be the only one.

Moreover, in favor of a purposeful approach says the fact that if you initially go from goals to resources, then the set of acceptable "rational" structures of s^{-1} is significantly narrower than the set of "rational" structures of s in the approach from resources to goals.

Thus, with the purposeful approach, the developer can proceed not from limited resources, but from a variety of tasks, goals, functions prescribed by the system.

With the purposeful approach, the main out of set $C = \bigcup_a C_a$ of subsets C_a of goals is a subset C_{af} and its corresponding functional architecture a_f, responsible for the implementation of the goals (functions) of the system for its purpose. The synthesis (selection) of the a_f functional architecture is explained by the commutative diagrams shown in Fig. 3a, b.

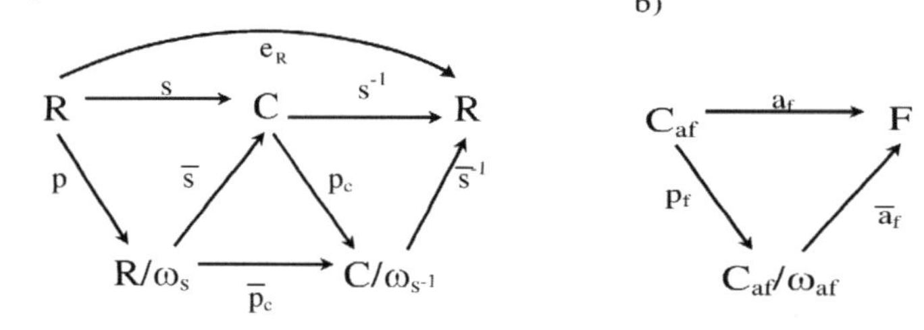

Figure 3. The selection of a functional architecture of a system structure.

Allocation from the General structure and Autonomous consideration of any architecture, as already shown, are possible due to the factorization of the structure due to the presence of independent strata (classes) of resources and the induction of their respective independent architectures in the structure of the system.

Indeed, from the proven statements follows bijectivity $\overline{p_c}$, that is, there is a one-to-one correspondence between the classes of resources R/ω_s and goals C/ω_{s-1}, which allows to solve the problem of synthesis of the system separately for each resource class and the corresponding target class.

The further decomposition of the system structure synthesis procedure is related to the factorization of the image $\overline{p_c}$ or the inverse image $\overline{p_c}^{-1}$, the narrowing of which a_f for subset $C_{af} \subset C/\omega_{s-1}$ is shown in Fig. 3b.

Factorization of functional architecture $a_f = \overline{a_f} p_f$ means grouping of sub-goals and corresponding functional resources into equivalence classes at a lower level of detailing, which provides outlining of the basic functional subsystems of the designed system. Such a splitting C_{af}/ω_{af} contains a subset of goals, providing solutions to the groups of system homogeneous tasks (such as information processing, control synthesis, etc.) and requires for its implementation a certain kind of resources from class (variety) of functional resources $F \subset R$ (belonging to the entire set of resources R) involved to create system. This consistent and deeper factorization of target sets into nested non-overlapping subsets allows you to build a target tree and the corresponding resource tree. At the stage of system or, as it was called earlier, external design [17], the depth of detailing of the goals, as a rule, is limited to the construction of a set of independent architectures that form the structure of the system. Subsystems combined into these architectures are considered as indivisible elements with corresponding connections. Further detailing of architectures and their forming elements-subsystems is carried out at the stage of schematic or internal design [17].

Along with the satisfaction to an external purpose at the present time the main is the quality or rationality of structure $\overline{s}$ of system $\sum = (R/\omega_s, \overline{s}, C)$ and the components of the structure of architectures (see Fig. 3) characterized by qualitative indicators (properties) and numerical criteria. These properties and criteria are analyzed independently for any of architectures $a \in s$ (by local properties and comparison criteria) and for structure s as a whole (by global properties and comparison criteria). It should also be borne in mind that in the structure of the system the depth of integration of elements in its constituent architectures may be different: due to the independence and uncorrelated architectures, some of them may be more integrated (strongly connected) than others.

3. Conclusion

Thus, it is shown that the purposeful approach, which has a clear meaningful sense and logical substantiation, allows both convenient mathematical description and substantiation using the language of modern algebra.

References

[1] Kalman R, Falb P and Arbib M 1971 *Essays on mathematical systems theory* (Moscow: Mir)
[2] Krasovsky A A 1987 *A handbook on the theory of automatic control* (Moscow: Nauka)
[3] Krasnova S A, Utkin V A 2006 *Cascade synthesis of observers of dynamic systems state* (Moscow: Nauka)
[4] Shahtarin B I 2008 *Wiener and Kalman filters* (Moscow, Gelios ARV)
[5] Luenberger D B 1966 Observers of multivariable systems *IEEE Trans* **AC-11** 190–197.
[6] Chernodarov A V 2017 *Control, diagnostics and identification of aircraft devices and measuring and computing complexes* (Moscow: "Nauchtechlitizdat" LLC)
[7] Isidori A 1995 *Nonlinear control systems* (Berlin: Springer-Verlag)
[8] Ciccarella G, Dalla Mora M and German A A 1993 Luenberger-like observer for nonlinear systems *Int. J. Control* **57**(3) 537–556.
[9] Proychev Ph and Mishkov R L 1993 Transformation of Nonlinear Systems in Observer Canonical Form With Reduced Dependency on Derivatives of the Input *Automatica* **29**(2) 495–498.

[10] Shouse K R and Taylor D G 1993 Discrete time observer for singularly pertubed continuous time systems *IEEE Trans. Automat. Control* **29**(2) 95–98
[11] Kulabukhov V S 2015 Aviakosmicheskoye priborostroyeniye 12:11–31. [Rus]
[12] Bukov V N, Kulabukhov V S, Maksimenko I M and Ryabchenko V N 1999 *Automation and Remote Control* **8** 61–73.
[13] Kulabukhov V S 2018 *IOP Conference Series: Materials Science and Engineering* **312** 012016 (https://doi.org/10.1088/1757-899X/312/1/012016)
[14] Kulabukhov V 2017 *MATEC Web of Conf.* **99** 03008 (doi: 10.1051/matecconf/20179903008)
[15] Kulabukhov V S 2018 The General principle of isomorphism in systems theory *Cloud of Science* **5**(3) 400-472 (https://cloudofscience.ru/sites/default/files/pdf/CoS_5_400.pdf) [Rus]
[16] Kostrikin A I 1994 *Vvedeniye v algebru. Osnovy algebry* (Moscow: Fizmatlit) [In Rus]
[17] Nikolaev V I and Bruk V M 1985 *Skhemotekhnika: metody i prilozheniya* (Mashinostroenie) [Rus]

IOP Conf. Series: Materials Science and Engineering **476** (2019) 012018 doi:10.1088/1757-899X/476/1/012018

The System of Uniformity of Measurement Assurance Facing Digital Transformation of the Economy

A Yu Kuzin and A V Yashin

All-Russian Research Institute of Metrological Service, Moscow 119361, Russia

E-mail: yashin@vniims.ru

Abstract. The article presents the guiding directions for the development of the system of the uniformity of measurement assurance in the Russian Federation. It addresses the challenges that can arise in connection with the implementation of the Digital Economy concept. The directions considered in the article are provided with the examples, the realization of which can be seen already today or will be seen in the nearest future.

1. Introduction

The term "digital economy" is generally understood as economic activities supported by digital information technology. The digital format of data, a key factor of production in all areas of socio-economic activities, makes the country more competitive, improves the quality of people's life, offers better conditions for doing business and, ultimately, ensures the economic growth and national sovereignty.

The challenging digital issues in Russia are presented in the documents of strategic planning, issued by the Russian leadership, the government's Program "Digital Economy of Russia" is among them. It must become one of the drivers accelerating scientific and technological development in the country. Many experts consider that in the implementation of this Program the active role should be played by enterprises of the aviation industry. This will benefit their competitiveness not only on the domestic market, but on the world market as well. The plan for overall digitalization is part of the state program "Development of the Aviation Industry Until 2025".

The digital economy will cause digital transformation of the system of uniformity of measurement assurance (SUMA), metrological support being its one of the most important elements. It is connected with the implementation of new technological capabilities in the industry – machine learning and artificial intelligence; processing of very large amounts of data, processing of unstructured data, real-time data processing; predictive management; a great number of cheap sensors and measuring devices; real-time control systems; augmented and virtual reality, etc. as well as new requirements for the system of uniformity of measurement assurance [1–3].

According to Russian metrologists, one of the results of the digital transformation will be a significant increase in the amount of sensors, measuring instruments (MIs) and measuring systems. This is one of the main challenges of today's metrology. In addition, these sensors, in most cases very tiny, will be used far and wide, even in difficult-to-reach places, and may be in possession of different owners. So, all this requires completely new approaches to be developed and applied towards metrological support, including remote automatic verification and calibration, the widespread use of the cloud technologies, and formation of large datasets (Big Data).

To adapt to new conditions new rules are needed and it will be necessary to develop and use normative documents regulating entirely new metrological procedures.

What steps are supposed to be taken today at the beginning of this not so easy path towards digital transformation of the UMA system?

To answer this question we will consider three main interrelated aspects of the digital transformation of the UMA system: legal framework, methodological, and technical.

2. Description of the research

2.1. The legal aspect

The legal aspect envisages a revision of the legal and regulation framework in the country, related to the uniformity of measurement assurance.

The digital economy management is based on forecasting the development of the digital economy and digital technology, on reviewing the monitoring results and controlling the fulfilment of the program for the development of the digital economy. Regarding the system of uniformity of measurement assurance, it means that its development should be based on forecasting the demand in measurements, monitoring the current state and coordination of development and production of measuring equipment in all sectors of the economy. Thus, the long-term tasks for the system of uniformity of measurement assurance in the Russian Federation are as follows:

- monitoring the system of uniformity of measurement assurance, forecasting the needs of the economy and society in measurements. The results of fulfilment of this task are important for the development of the economy as a whole, and are to be available to a vast majority of legal entities and physical persons, and individual entrepreneurs;
- coordination of the activities of the authorities in the development and production of measuring equipment in the country. This important activity focuses on satisfying the requirements of all sectors of the economy in measuring instruments (for example, intelligent sensors), and is the key condition for the digitalization of production through introduction of internet of things in the industry.

These important for the digital economy tasks are to be fixed in the national law regulating the uniformity of measurement assurance.

An important element of the digital transformation of the system of uniformity of measurement assurance is the application of electronic document management system, a tool which allows to radically improve the interaction between various entities in the digital environment. The electronic document management is especially important for the procedures, directly connected with production, trade, health services, etc., such as type approval of measuring instruments (reference materials), and verification of measuring instruments. These procedures are used to determine the legal use of measuring instruments (reference materials) in the area of state regulation.

Today's practice of document management, used for metrological procedures for the purpose of type approval of reference materials (RMs) or measuring instruments (MIs), as well as verification of MIs, the results of which determine the applicability of MIs and RMs in the state-regulated sphere, is based upon paperwork. At the same time there is a procedure for downloading the results of these metrological procedures into the State Information System. However, this information is not legally bound. Therefore, such information cannot be used by the federal state metrological supervision bodies as references for their evidential basis, or it cannot be used by legal entities and individual entrepreneurs as evidence that the MI they want to use is of an approved type and complies with metrological requirements. The country has at its disposal huge amounts of information related to the uniformity of measurement assurance, however their practical value is limited. This provides a good reason to develop proposals for the transition from paperwork to electronic registration of the results of work (services) concerning type approval of the MI (RM) or verification of the MI and thus giving this information legal relevancy.

Digital transformation in the uniformity of measurement assurance, as other countries' experience has shown, inevitably results in building spatially distributed measuring systems with numerous intelligent sensors in them and cloud structures for storing and processing data, "digital twins" of measuring instruments to be integrated in the automated technological process control systems, "digital twins" of measurement standards to perform remote verification (calibration), etc. All these call for developing new requirements for measurements, RMs and MIs. In particular, the simulation of measurement process of real parameters of objects during virtual testing will be legally regulated, and also the methods for virtual measurement of parameters of objects will be prescribed, which will require appropriate revision of the national legislation related to the uniformity of measurement assurance.

2.2. The scientific and methodological direction

The scientific and methodological direction is primarily connected with the solution of the above mentioned task of monitoring the UMA system, forecasting the demand of the economy and society in measurements.

Investigating this issue, specialists have developed a concept of monitoring and predictive assessment of measurement needs, which is summarized here as the following sequence of steps:

1. Gathering initial information on measurement requirements by conducting surveys among federal executive authorities, industrial enterprises and companies, state regional metrology centers. Information is to be submitted in a standardized data format. It should be mentioned that the collecting of information is already underway.

2. Summarizing and processing the gathered information. Afterwards a multidimensional array of data is formed, giving an idea of the level of satisfaction of current needs and the projected demand in measurements for a certain time frame, the ratio of the amount of calibration to verification work, the characteristic of such work (services) as testing, verification, etc. This information can be compiles as follows:

- federal executive bodies;
- branches of industry;
- areas of activities;
- spheres of state regulation in assurance of measurement uniformity;
- measurement domains.

The acquiring of data and reviewing the results of the monitoring and forecasting of the requirements in measurements allow to identify key factors in realizing the goals and solving the tasks of the development of the UMA system, and also in achieving the targeted values of development indicators within the managing time-frame.

3. Based on the monitoring results and forecasting, the development of proposals concerning the creation of measurement standards.

4. Based on the information acquired while implementing steps 2 and 3, taking efficient management decisions for the benefit of the UMA system.

It is evident that today the best effect of the realization of the proposed concept can be achieved through a software-distributed system, the stakeholders (subscribers) of which operate in a single information and communication space. However, the concept of monitoring and forecasting the requirements in measurements can be realized through software only elementwise. To this end, a prototype of a hardware-software system (HSS) has been developed, providing the automated solution of problems described at stages 3 and 4. This HSS implements modern mathematical methods for information processing and modeling of, for example, genetic algorithms used to solve problems of optimization. The use of an HSS prototype can help getting options for optimizing work plans for the development, modernization and maintenance of state primary standards within a given time-period of planning, with various scenarios of funding.

2.3. The technical direction

The technical direction is associated today with meeting such challenges as the development of an e-platform to be used for sharing information by manufacturers, suppliers, and consumers; the simulation (modeling) of measurement processes for real-world parameters of objects during virtual testing; the development and implementation of a digital calibration certificate; methods (procedures) for the virtual measurement of certain parameters of an object, etc.

Calibration of measuring instruments is one of metrological activities. The calibration results are certified by issuing a calibration certificate. Introduction of the digital calibration certificate is being widely discussed today in many countries. The document ISO/IEC 17025 "General requirements for the competence of testing and calibration laboratories" contains minimum requirements for the calibration certificate. It envisages a printed on paper version of the certificate and an electronic version, which is in fact an electronically made copy of the paper certificate. Meanwhile, we consider the concept of "a digital calibration certificate" needs to be expanded.

The digital calibration certificate helps solve the key task – the confirmation of metrological traceability of the measurement result, allowing the result to be related to a reference through a documented unbroken chain of calibrations, each contributing to the measurement uncertainty. For the "reference" can be a definition of a measurement unit through its practical realization, or a measurement procedure, or a measurement standard.

Metrological traceability requires an established calibration hierarchy, i.e. a sequence of calibrations, starting from the basis for comparison and ending with a working measuring instrument. What is more, the result of each calibration depends on the result of the previous one in this sequence. The digital calibration certificate must allow to monitor each step of the calibration process with maximum credibility, taking into account all the necessary data that ensure the obtaining of the specified calibration result. Meanwhile no access is permitted to the chain of calibrations to make changes. Provided any calibration step is not available or replaced by some fictitious data the impossibility of obtaining a digital calibration certificate should be guaranteed. Thus, the consumer will be protected from sham or even fraudulent activities. A digital calibration certificate that meets the above-mentioned requirements can be obtained through block-chain technology.

Metrological traceability does not guarantee that the measurement uncertainty corresponds to the measurement task being solved. In this case, the consumer analyzing the calibration chain should be able to determine the stage of the hierarchical calibration chain that will satisfy him/her in terms of uncertainty or some other indicators. Our understanding of the digital calibration certificate is that it is a software product that provides the consumer with a number of services, enabling a successful solution of a measurement task with the help of a calibrated measuring instrument.

Another priority task that needs to be solved is the development of an e-platform to be used for sharing information among manufacturers, suppliers and consumers of measuring equipment. The need for such platform is caused by the absence of the information base today, that could accumulate the consumers' needs in measuring instruments and the capabilities of the industry to meet them. Because of this manufacturers are not able to form a long-term package of orders that could justify the financial investments necessary for the production of measuring instruments, and consumers, in their turn, do not know who should they address for a prompt solution of their problems. The suggested e-platform, in our opinion, should exercise the following functions:

- information: providing access to view the list of customers (organizations) and consumers (organizations), working on the e-platform, obtaining information regarding any organization of interest;
- marketing: providing information concerning MIs which are in demand, and their potential customers; helping choose designers, manufacturers, suppliers of the required measuring instruments;
- advertising: making available in a single information space the data regarding an organization by placing by that organization their own information on the e-platform;

- trade: allowing to organize competitive bidding, and carry out a whole range of various trade and procurement measures for the acquisition and execution of work (services) associated with the development, manufacturing, and supply of measuring instruments;
- analytical: allowing to evaluate the activities of the organizations registered on the e-platform by various indicators, to compare the measuring instruments offered, to make a choice of partners;
- security of information: ensuring the security of electronic document management with the help of certified cryptographic information security tools.

Such e-platform will allow to respond with maximum flexibility to the market needs in domestically manufactured measuring equipment, will provide MI manufacturers involved in high-tech areas a gradual transition from making one-time orders to making long-term production programs, which will reduce the cost of domestically manufactured MIs.

3.Conclusion

In conclusion, we would like to note that the above described directions of digital transformation of the UMA system form the basis for realizing a set of measures to ensure the agreement of the UMA system with the tasks of development of the digital economy, focusing both on the development of the existing conditions and creating new ones for the emergence of promising digital technologies in the assurance of measurement uniformity.

References

[1] Hackel S, Härtig F, Hornig J and Wiedenhöfer T 2017 The Digital Calibration Certificate *Metrologie für die Digitalisierung von Wirtschaft und Gesellschaft*, pp 75-82 (DOI 10.7795/310.20170499)

[2] PTB-Digitalisierungsstudie 2017 Metrology for the digitalization of economy and society, *PTB-Mitteilungen* **4** (DOI 10.7795/310.20170401EN)

[3] Thiel F 2018 Digital transformation of legal metrology - The European Metrology Cloud *OIML Bulletin* **59**(1) 10-21

IOP Conf. Series: Materials Science and Engineering **476** (2019) 012019 doi:10.1088/1757-899X/476/1/012019

Developing a Method for Exploring the Cyclic Deformation of Textile Materials for Interior Finish of Aircraft Cabins

L N Lisienkova[1], I A Savenkova[2] and E V Baranova[3]

[1]Moscow Polytechnic University, 38, B. Semenovskaya street, 107023, Moscow
[2]National Research University Higher School of Economics, 20, Myasnitskaya str., 101000, Moscow
[3]National Research University (branch) South-Ural State University, Zlatoust

E-mail: lisienkovaln@mail.ru

Abstract. The paper discloses the method for estimating the deformation of fibroreticulate materials under the conditions of spatial multiaxial cyclic tension. The relevance of the method application for estimating the reliability performance of materials used for upholstering and finishing the interior of aircraft cabins has been justified. The equations of the elastic state of flexible fibroreticulate materials are obtained under the spatial tension in stresses and deformations. The geometric parameters of elastic deformations in the material during in the material in the design, processing treatment (molding process) and operation (shape stability) of the products. The advantage of the developed method of spatial cyclic tension of flexible fibroreticulate materials is the ability to model test conditions that simulate operating conditions. This capability of the test method allows to examine the dynamics of changes in the markers of the molding properties and shape stability of materials and to predict their behavior during operation.

1. Introduction

The assortment of textile materials and products for the aircraft industry and aeronautics is constantly expanding and updating. Textile fabrics for aircrafts are widely used for upholstering the interior of an aircraft, fuel tanks of modern fighters, wings and tail of single-engine airplanes, for manufacturing brake parachute canopies, covers for storing aircraft equipment and frame covering. In recent times, marketing competition makes the air companies give pride of place to the selection of materials for interior finish of aircraft cabins.

The interior of the aircraft cabin is a system of specifically developed and artistically designed elements of the aircraft interior, which provide aesthetic perception and favorable living environment during the flight, improve safety of passengers in case of emergency. The structural elements of the interior include:

- seat covers (seat, back, armrests, head rests);
- seat cushions (seat, back, armrests, head rests);
- carpet mat;
- curtains and so on.

To upholster and finish the interior of an aircraft cabin, a wide range of fibroreticulate materials with various structure and high strength are used:

- cloth;
- knitted fabric;
- natural leather;
- synthetic leather;
- fur;
- foam for seat cushions;
- carpet mat;

It should be noted that the passenger spends 90% of the flight time alone, surrounded by the elements of the interior, the state of which creates a certain mood, depending on the state of the interior. Clean new carpets and covers, comfortable back cushions and seats make a flight very comfortable. Shabby carpets, out-of-shape covers, bowed cushions lead to fatigue and irritation, and the passenger has a negative impression of the service quality of an air company.

During manufacturing and operation, the materials of the interior products of aircraft cabins (covers for seat, pillows, head rests, curtains) are mainly subjected to cyclic force. Mechanical force on the materials during manufacturing and operation of these products amounts to no more than 10–30% of the material strength. The main mechanical factors that change the deformation properties of the materials and products during manufacturing and operation are tensile and compressive forces. External loads lead to complex spatial deformation of the structural elements of fibroreticulate materials, changing the size and shape of parts and products of the cabin interior.

Therewith, the limiting characteristics of the strength change insignificantly and reflect the nature of structural changes in the materials during technological and operational impact inadequately. The phenomena of wear and service life of modern textile materials are mainly associated with the accumulation of non-recoverable deformations in the materials due to the macrostructure changes.

The analysis of the published works has shown that the applied methods for estimating the deformation indicators of: stiffness, crease retention, elasticity, shrinkage and others are suitable for controlling the initial quality of the materials.

Standard or mainstream devices for the material testing for repeated tension have not yet been developed. There are individual device models used in industry, research laboratories and educational institutions. They work on the principle of preserving a constant amplitude of absolute deformation, a constant amplitude of relative deformation, a constant amplitude of cyclic load in each cycle (МР-2, РП, УП-1, МТИ, ПД-5М, МРД-1, etc.). [1]. The analysis of the construction solution of these devices shows that basically they have a mechanical principle of the sample loading and a visual measurement of the deformation, which leads to significant errors in the measurement results.

In addition, almost all of these methods are based on cutting the samples with subsequent long-term tests of the latter using various laboratory instruments, which makes them rather material-intensive and time-consuming.

It should be noted that the most advanced method is the new method for estimating the deformation properties of the leather based on computer analysis of the material relaxation process after the unloading. The major disadvantage of the device is that the indenter is moved by the mechanical impact of a person, which can cause errors during the testing [2].

The major disadvantage is that the existing methods and tools do not adequately reflect the influence of structural changes in the materials during technological and operational impact on their deformation properties. This fact does not provide an adequate assessment of the technology indices and reliability of the materials and products.

The problem resides in the absence of methods and tools to assess the impact of structural changes in the materials on their deformation properties adequately. This problem does not allow choosing various modern materials for upholstering and finishing the interior of the aircraft cabins efficiently and predicting their behavior during operation.

The purpose of the paper is to develop methods and tools for testing the materials under cyclic spatial tension.

For the purpose in view and scientific justification of the proposed method, it has been necessary to conduct the analysis of the material deformation under spatial tension.

2. Analysis of the material deformation under spatial tension

The paper considers the deformation of a flexible shell with perpendicular applied force $\overline{P}$. To calculate the area of the deformed object of research, it is assumed that the deformable material is an elastic homogeneous shell (Fig. 1) that has a solid cross section of a small initial thickness $L_0$and volume V_0. For simplicity, the circuit to fix the shell of radius R_1 and friction of the material surfaces and the loading element have not been taken into account.

The change in the shell shape occurs as a result of moving a point that is at the peak to the distance of h_{max} (Fig. 1.). At the fixed position of the shell along the circuit *AB*, its movement occurs due to the thickness loss without change of the volume. To find the surface area of the shell under tension, it is necessary to find the areas of the figures (1) and (2) – S_1 and S_2 respectively. To do this, it is necessary to determine the indenter base radiusr_1 and establish the interconnection between the movement of *H* and the element of cylinder of the deformable figure *l*.

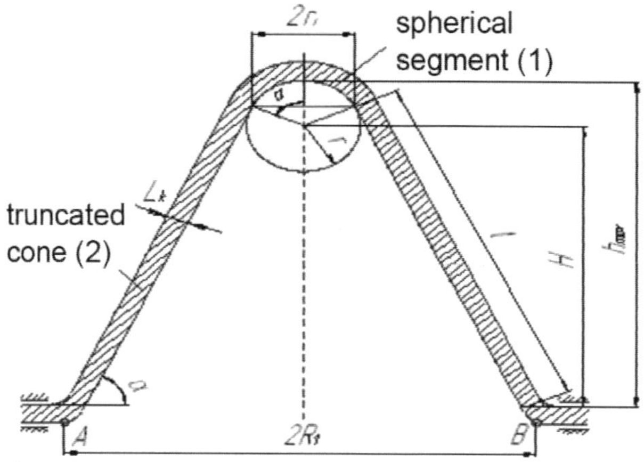

Figure. 1. Spatial deformation scheme.

By algebraic transformations, a solution to the equation for the indenter base radius r_1 is found:

$$r_1 = \frac{Hr\sqrt{H^2 + R_1^2 - r^2} + r^2 R_1}{H^2 + R_1^2}. \tag{1}$$

Next, the areas S_1, S_2 of the figures (1) and (2) are found geometrically:

$$S_1 = \pi r_1^2 + 2\pi r \cdot \left(r - \sqrt{r^2 - r_1^2} \right). \tag{2}$$

$$S_2 = \pi \left[R_1 \sqrt{R_1^2 + h^2 \left(1 + \frac{r_1}{R_1 - r_1} \right)^2} - r_1 \sqrt{r_1^2 + \frac{h^2 r_1^2}{(R_1 - r_1)^2}} \right], \tag{3}$$

where $h = H + \sqrt{r^2 - r_1^2}$.

Analysis of equations (2) and (3) and their experimental estimate has shown that the material surface area under spatial tension depends on the sample R_1, indenterr_1 and its movement value *h*, the correlation coefficient between these values has been 0.86 ... 0.92.

The obtained formulas (2) and (3) as applied to the real objects allow us to calculate the internal surface area of the clothes objectively. Therefore, to correct the above equations (2) and (3), the experimental estimate of the change in the material thickness under spatial tension has been carried out.

The impact of the material thickness on the error in estimating the value of the material deformation under tension has been studied experimentally. As a result, the correction factor in the formula (3) has been determined, allowing to exclude the error in the deformation estimate for the materials with a thickness of more than 2.0 mm. It has been established that the equation (9) calculates the sample surface area of the material with a thickness of 0.01 mm to 2.0 mm with an error of 2.8–10.3%.

The papers [3–6] and other studies have shown that the problem to find an adequate model of the equilibrium mode of deformation of the researchable objects under spatial tension is related to the specifics of their structure and properties, including anisotropy and porosity.

Fibroreticulate polymeric materials, unlike the perfect flexible shells, can change not only the thickness but also the volume during deformation. The experience of theoretical modelling of the spatial deformation of such materials ignoring their properties leads to a significant error in the known models and limitation of their adequacy during the real process. Therefore, the modelling has involved the consideration of the material capability to change volume during deformation.

Conditionally, the shell sections can be represented as a sphere (hemisphere, cone), on the inner surface of which the pressure P, Pa (Fig. 2) is distributed.

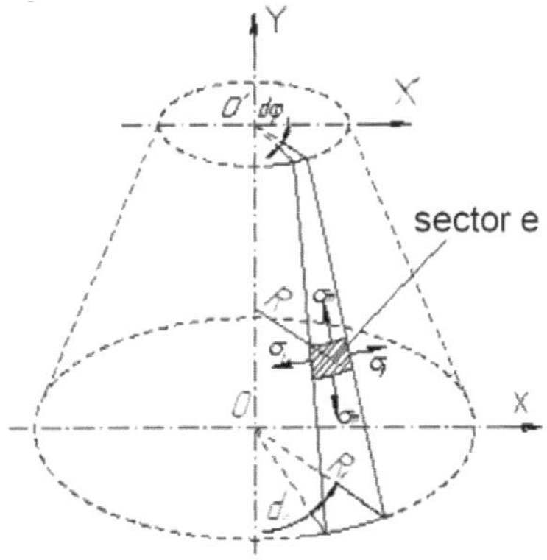

Figure. 2. Scheme to the analysis of stresses in the shell under tension.

In this case, the well-known Laplace equation is used:

$$\frac{\sigma_m}{R_m} + \frac{\sigma_t}{R_t} = \frac{P}{L_0}, \tag{4}$$

where σ_m – meridian stress, Pa; σ_t – parallel (tangential) stress, Pa; R_m – radius of curvature in the meridian, mm; R_t – radius of curvature about the axis, mm; L_0 – shell thickness, mm.

For real conditions of tension, the full symmetry of the shell (as it is often assumed in such calculations) will be only a special case of deformation of anisotropic materials. Therefore, radius of curvature in the meridian R_m and about the axis of the figure R_t will differ in most cases. If we assume that $R_m = \infty$, $R_t = R$ (Fig. 2), the task to find the deformations and stresses in the shell is reduced to two-dimensional model and, therefore:

$$\frac{\sigma_t}{R} = \frac{P}{L_0}. \tag{5}$$

In this case, to estimate stresses, it is necessary to know the initial thickness of the shell. Provided that the shell volume is preserved $V_к = V_0$, the finite thickness of the shell $L_к$:

$$L_к = \frac{V_к}{S_к} = \frac{\pi R_1^2 L_0}{S_1 + S_2}. \tag{6}$$

However, for the real objects, the condition of preserving their volume under tension is not satisfied; therefore, the correction coefficient $k< 1$, ($V_к = kV_0$) has been included in the equation (6). Then, the relative deformation is as follows:

$$\varepsilon = \frac{L_0 - L_к}{L_0} = 1 - \frac{k \cdot L_к}{L_0} = \varepsilon_m + \varepsilon_\tau . \quad (7)$$

For anisotropic flexible materials, the interconnection between stress and deformation under tension is non-linear. But within the limits of elasticity for the elementary section of the shell e (Fig. 2), it is possible to write according to the Hook's law:

$$\left.\begin{aligned} \varepsilon_m &= \frac{1}{E}(\sigma_m - \mu\sigma_\tau) \\ \varepsilon_\tau &= \frac{1}{E}(\sigma_\tau - \mu\sigma_m) \end{aligned}\right\} , \quad (8)$$

where E – the elasticity modulus of the 1st kind (Young modulus) under tension, Pa; σ_m, σ_τ – meridian and parallel (tangential) stresses, Pa; μ – the lateral contraction ratio (Poisson's ratio).

Having solved the system of equations (8), we have determined the meridian deformation ε_m:

$$\varepsilon_m = \frac{\sqrt{\left(H + \sqrt{r^2 - r_1^2}\right)^2 + \left(R_1 - r_1\right)^2} + r \cdot \arccos\frac{r_1}{r}}{R_1} - 1 . \quad (9)$$

Parallel deformations ε_τ are found from the equation (7) by the substitution of ε_m.

Under the condition of homogeneity of the shell cross-section, the average values of σ_m and σ_τ, are:

$$\begin{cases} E \cdot \varepsilon_m = \sigma_m - \mu\sigma_\tau \\ E \cdot \varepsilon_\tau = \sigma_\tau - \mu\sigma_m \end{cases} \quad (10)$$

Meridian and parallel stresses are determined by solving the system (10)

$$\sigma_\tau = \frac{E\left(\mu\varepsilon_m + \varepsilon_\tau\right)}{1 - \mu^2} . \quad (11)$$

$$\sigma_m = \frac{E\left(\mu\varepsilon_\tau + \varepsilon_m\right)}{1 - \mu^2} . \quad (12)$$

To solve the equations (8), (11), (12), the Poisson's ratio and the modulus of the 1st kind (elasticity) under tension E are used, which are determined experimentally or specified, taking into consideration the purpose function.

The developed deformation equations allow estimating the elastic deformations and the resulting stresses in the material under the influence of external force actions. On the one hand, the equations (11) and (12) allow predicting the force that should be applied to the material during molding (molding ability). On the other hand, the specified equations allow determining the force, at which elastic deformations occur in the object and its shape and size (shape stability) are preserved.

The comparative analysis of the experimental results of the material testing using the developed device with the data obtained theoretically [7] has confirmed the adequacy of the developed model within the elastic deformation.

3. Developing a method for assessing the deformation of fibroreticulate materials under spatial tension

To carry out the tests, a method and a device for cyclic spatial tension have been developed (RF Patent No. 2354953) [8], the fundamental difference of which lies in the possibility to measure the deformation components of material samples after the specified number of cycles of their spatial tension.

It allows us to study the dynamics of changes in the deformation indicators of materials under the conditions of the sample testing that simulate the impact of technological and operational factors.

A method for estimating the deformation of flexible materials is proposed, based on the use of analytical-type tool, which makes it possible to implement a comprehensive and more accurate control of all input and output parameters.

The principle of operation of this device lies in the fact that the test material is fixed along the ring circuit and is loaded by the indenter using an automatically specified multi-cycle mechanical force.

Therewith, the material is subjected to multi-axial spatial tension, obtaining equal deformations along all elements of the cone. The load on the indenter is integral from the loads in all directions and, therefore, it is independent of the leather position relative to the device. Thus, during the extrusion test of the sample clamped along the ring circuit, the average characteristics of the anisotropic material are obtained, regardless of its position.

In the proposed device, the movement of the indenter after the unloading is converted by means of a linear increment encoder into an electrical signal and transmitted to the computer in real time digitally. Further, the obtained information is processed by a special program that calculates the main deformation characteristics.

The proposed method allows measuring the deformation directly in semi-finished or finished products. The device has easy design and is simple to operate. At the same time, it allows to obtain more accurate and objective measurement results.

The subject matter of the method consists in estimating the complete, reversible, irreversible deformation component (in % and fractions), the calculation of which uses the measurement data of the material sample deflection under the loading h_{max}, after the unloading h_1 and after the rest h_{rest} (Fig. 3).

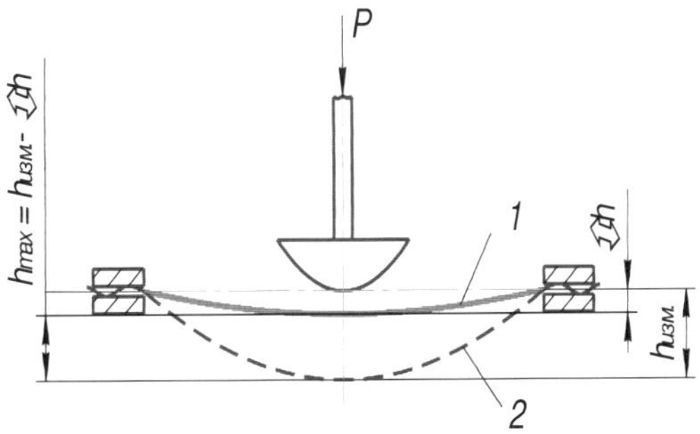

Figure. 3. The position of the sample material before testing (1) and in the conditions of spatial tension (2)

The measuring system of the device (Figs. 4, no. 1–9) allows to: automate the control and operation of all mechanisms of the device; automatically measure and consider the value of the initial sample deflection relative to the zero level Δh under the current measurements (see Fig. 14); simulate test conditions. High accuracy of the measurement results of the sample deflection value (instrument measurement error is not more than 10 μm) is provided by differential photosensors (see. Figs. 4, no. 5–6).

The number of papers, including [8, 9], have experimentally determined the optimal parameters for testing fibroreticulate materials of different fibrous composition and structure: the time of sample loading/rest in a cycle is 5–20/5–10 s; cyclic load on samples of fabric, leather and packages is 1.0...2.0 daN, on knitted fabrics is 0.5 daN.

The dynamics of changes in the properties of materials under the influence of external factors is analysed on the basis of determining the indicators of plasticity (%), elasticity (%), conditional stiffness (N). The calculation of the relative deformation (%), average stress σ (MPa), the equilibrium elastic modulus E_3 (MPa) is performed using the equations (7) - (11) in the MathCAD system.

The paper [9] and other studies have performed the experimental verification of the theoretically developed method of spatial deformation during the study of samples of leather, fabrics, knitted fabrics, as well as packages obtained by molding the parts of fabrics and leather by means of doubling and processing with polymer dispersions.

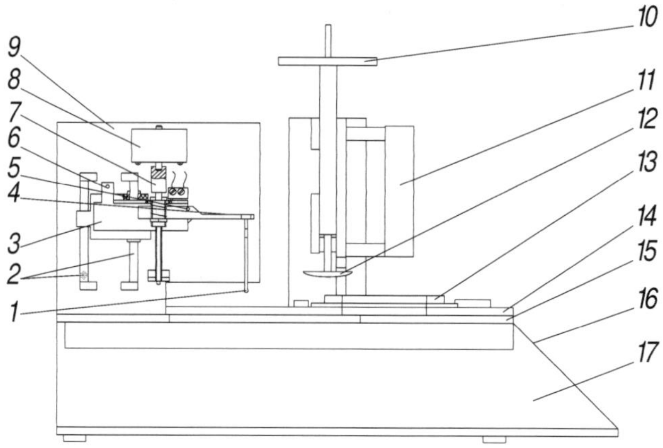

Figure. 4. The device scheme for cyclic spatial tension: measuring system (1–9), including: 1 – spring contact probe, 2 – antibacklash plug, 3 – guidance, 4 – balancing lever, 5 – probe photosensor, 6 – differential photosensor, 7 – carriage, 8 – lead screw, 9 – measuring device; 10 – the supporting point for the placement of goods; 11 – electric motor; 12 – loading device; 13 – clamping rings to fix the sample; 14 – carriage moving over the table 15; 16 – step motor; 17– electronic assembly with input panel and data display.

4. Conclusion

1. The equations of the elastic state of the flexible fibroreticulate materials are obtained under the spatial tension in stresses and deformations.

2. The geometric parameters of elastic deformations in the material during in the material in the design, processing treatment (molding process) and operation (shape stability) of the products.

3. The advantage of the developed method of spatial cyclic tension of flexible fibroreticulate materials is the ability to model test conditions that simulate operating conditions. This capability of the test method allows to examine the dynamics of changes in the markers of the molding properties and shape stability of materials and to predict their behaviour during operation.

References

[1] Zhiharev A 2004 *Material studies in light and clothing industry: study scriptum for institutions of higher education* (Moscow: Akademia)

[2] Zhikharev A 2005 *Development of scientific foundations and methods for evaluation the material quality for textile industry products under force, temperature and humidity influences* (PhD thesis)

[3] Belokurov V 2007 *Advancement of theoretical basis and development of methods for determining the viscoelasticity of consumer industry materials in quasistatic and dynamic resonant modes* (PhD thesis)

[4] Belyaev N 1965 *Resistance of materials* (Moscow: Nauka)

[5] Zybin A 1974 *Biaxial tension of materials for the shoe upper* (Lyogkaya Industriya)

[6] Makarov A 2005 *Development of computer technologies for modeling the physical and mechanical properties of textile materials with complex structure* (PhD thesis)

[7] Lisienkova L 2010 Theoretical analysis of the deformation condition of materials for clothing in conditions of spatial stretching *Universities bulletin*, **2** 28-31

[8] Lisienkova L, Baranova E, Stelmaschenko V and Salamatin A 2009 *Patent 2354953 S2 RossiyskayaFederatsiya, MPK G01N 3/08 Ustroystvo dlya opredeleniya deformatsionnykh svoystv kozhi i podobnykh yey gibkikh materialov*

[9] Lisienkova L and Smolina O A 2016 Experimental Studies of Deformation of Knitted Fabrics under Uniaxial and Cyclic Spatial Sprains *Universities bulletin* **3** 163-166

IOP Conf. Series: Materials Science and Engineering **476** (2019) 012020 doi:10.1088/1757-899X/476/1/012020

Nonlinearities Evaluation in the Aircraft Parameter Identification

S V Nikolaev[1] and O N Korsun[2]

1 Valery Chkalov State Flight Testing Center, Akhtubinsk, Russia
2 State Research Institute of Aviation Systems, Moscow, 125319, Russia

E-mail: marmotto@rambler.ru

Abstract. The authors consider the algorithmic and methodological support for estimating nonlinear dependencies when performing the parameter identification of the aircraft mathematic models using the flight test data. Examples of the considered methods practical application to identifying the nonlinear parameters of several modern aircrafts are presented.

1. Introduction

The work refers to the problem of algorithmic and methodological support for aircraft parameter identification from the flight tests data. Currently in this area a significant progress is achieved. Thus, the basic approaches and the main stages of the identification technology are described in [1-3], the identification algorithms and examples of practical use in [4-7]. However, the problem of improving identification methods still remains relevant. The identifying the aerodynamic coefficients of an aircraft is a complex task, the solution of which should apply a wide range of techniques: decomposing a general task into a number of simpler ones, assessing the reliability of results by comparing data from different sources, using reliable results from related disciplines. The paper discusses the use of these techniques for solving one of the most difficult tasks of processing flight experiment data, which include the identification of nonlinear aerodynamic dependencies. At the same time, much attention is paid to the form of presentation of results, since in the task of identification, taking into account the characteristics of the operator and convenience of perception of the results significantly increase the interpretation.

2. Mathematical model of spatial motion

In the problem of identifying the aircraft aerodynamic coefficients, the most reliable result is the spatial motion equations, obtained directly from the basic laws of mechanics. These differential equations form the following non-linear system, where it is assumed that the aircraft has a symmetry plane[8]:

$$\frac{d\alpha}{dt} = \omega_z - \frac{1}{\cos\beta}\left[\left(\frac{a_x}{V} - \omega_y \sin\beta\right)\sin\alpha + \left(\frac{a_y}{V} + \omega_x \sin\beta\right)\cos\alpha\right],$$

$$\frac{d\beta}{dt} = \frac{a_z}{V}\cos\beta - \left(\frac{a_x}{V}\sin\beta - \omega_y\right)\cos\alpha + \left(\frac{a_y}{V}\sin\beta + \omega_x\right)\sin\alpha,$$

$$\frac{dV}{dt} = a_x \cos\alpha \cos\beta - a_y \sin\alpha \cos\beta + a_z \sin\beta,$$

$$\frac{d\omega_z}{dt} = \frac{I_x - I_y}{I_z}\omega_x\omega_y + q\frac{S\, b_A}{I_z}m_z - \frac{k_d\omega_y}{I_z} - \frac{(P_{right} + P_{left})\,y_d}{I_z} + \frac{I_{xy}(\omega_x^2 - \omega_y^2)}{I_z}$$

$$\frac{d\omega_y}{dt} = \frac{I_x(I_z - I_x) - I_{xy}^2}{I_x I_y - I_{xy}^2}\omega_x\omega_z + \frac{k_d\omega_z I_x}{I_x I_y - I_{xy}^2} + \frac{(P_p - P_l)\,z_d I_x}{I_x I_y - I_{xy}^2} -$$

$$- \frac{I_{xy}(I_z - I_x - I_y)}{I_x I_y - I_{xy}^2}\omega_y\omega_z + \frac{I_x q S\, l m_y + I_{xy} q S\, l m_x}{I_x I_y - I_{xy}^2},$$

$$\frac{d\omega_x}{dt} = \frac{I_y(I_y - I_z) + I_{xy}^2}{I_x I_y - I_{xy}^2}\omega_y\omega_z - \frac{I_{xy}(I_y + I_x - I_z)}{I_x I_y - I_{xy}^2}\omega_x\omega_z +$$

$$+ \frac{I_y\, qS\, l m_x + I_{xy}\, qS\, l m_y}{I_x I_y - I_{xy}^2} + \frac{I_{xy}(k_d\omega_z + (P_{right} - P_{left})\,z_d)}{I_x I_y - I_{xy}^2}, \tag{1}$$

$$\frac{d\vartheta}{dt} = \omega_y \sin\gamma + \omega_z \cos\gamma,$$

$$\frac{d\gamma}{dt} = \omega_x - tg\,\vartheta\,(\omega_y \cos\gamma - \omega_z \sin\gamma),$$

$$\frac{d\psi}{dt} = \frac{1}{\cos\vartheta}(\omega_y \cos\gamma - \omega_z \sin\gamma),$$

$$\frac{dH}{dt} = V\left[\cos\alpha\, \cos\beta \sin\vartheta - \sin\alpha\, \cos\beta \cos\vartheta \cos\gamma - \sin\beta \cos\vartheta \sin\gamma\right]$$

acceleration along the body-fixed axes

$$a_x = \frac{qS(-c_x + c_p)}{m} - g\sin\upsilon,$$

$$a_y = \frac{qSc_y}{m} + g\cos\upsilon \sin\gamma, \tag{2}$$

$$a_z = \frac{qSc_z}{m} + g\cos\upsilon \sin\gamma$$

overloads along the body-fixed axes:

$$n_x = \frac{qS(-c_x + c_p)}{mg},\ n_y = \frac{qSc_y}{mg},\ n_z = \frac{qSc_z}{mg}, \tag{3}$$

where α, β – angles of attack and sideslip, rad; ω_x, ω_y, ω_z – angle rates about body-fixed axes frame, rad/s; υ, γ, ψ – angles of pitch, roll, yaw, rad; V – true air speed, m/s; H – flight altitude, m; m_x, m_y, m_z – coefficients of aerodynamic torques; c_x, c_y, c_z – coefficients of aerodynamic force in body-fixed coordinates system; I_x, I_y, I_z, I_{xy} – momentum of inertia about axes of body-fixed frame, kg·m^2; m – aircraft's weight, kg; l, b_A – wing span and length of mean aerodynamic chord, m; S – wing surface area, m^2; $q = \rho_H V^2/2$ – dynamic air pressure, Pa; ρ_H – air density at the altitude of flight, kg/m^3; $c_p = P/qS$ – thrust coefficient; P_{right}, P_{left} – thrust of right and left engines, N; k_d – kinematic momentum of engine rotors, kg·m^2/s; y_d, z_d – engine coordinates in body-fixed axes frame, m.

The system of differential equations (1), and algebraic relations (2) and (3), enables to construct a model to check the consistency of onboard measurements[1, 6] , as well as models to identify longitudinal and lateral movement [1-3].

3. Identification of aerodynamic characteristics

Identification of aerodynamic characteristics should be carried out separately for the longitudinal and lateral channels. The corresponding models can be easily obtained from the general model (1) with regard to formulas (2) and (3) in the same way as in [1, 6] the model is obtained for checking the flight data. To identify the lift coefficients and the pitch moment, it's enough to take the equations for the angle of attack, the angular velocity of pitch as the model of the object, and in the model of observations to use the angle of attack, angular velocity and normal overload. To identify the lateral movement coefficients, the model for the object must include equations for the sideslip angle and angular velocities of the roll and yaw. The observation vector contains the measured values of the same signals, that is, full-component observations, as well as lateral overload. When identifying a longitudinal channel, the variables of lateral movement are replaced by measurements [5], and vice versa.

Aerodynamic characteristics are usually non-linear. The generally accepted approach is to use the data partitioning [1], that is to sort the data, for example, by the angle of attack: $\alpha \in [\alpha_k, \alpha_{k+1}]$, $k = 1,2,\ldots,N_k - 1$, where $[\alpha_k, \alpha_{k+1}]$ is the interval of the values of the angle of attack 2 ... 4 degrees wide; N_k - the number of sections splitting the full range of variation of the angle of attack. It is assumed that in this area the object is linear. Then, for example, the longitudinal channel coefficients may be presented

$$c_{ye} = c_{y0} + c_y^{\alpha}\alpha + c_y^{\delta_B}\,\delta_B,$$

$$m_z = m_{z0} + m_z^{\alpha}\alpha + m_z^{\delta_B}\,\delta_B + m_z^{\bar{\omega}_z}\cdot\frac{b_A}{V}\omega_z + m_z^{\bar{\dot\alpha}}\cdot\frac{b_A}{V}\frac{d\alpha}{dt}$$

However, the linearity assumption is not always true. Therefore, to approximate non-linear dependencies, polynomials or splines are used, the coefficients of which are included into the vector of identifiable parameters. The orders of polynomials or splines in this case are usually chosen not higher than the third. For example, the following approximations are used to estimate the dependence of the pitch moment coefficient on the angle of attack:

- polynomials of the second or third order

$$m_z(\alpha) = m_{z0} + m_z^{\alpha}\,\alpha + m_z^{\alpha 2}\,\alpha^2, \qquad m_z(\alpha) = m_{z0} + m_z^{\alpha}\,\alpha + m_z^{\alpha 2}\,\alpha^2 + m_z^{\alpha 3}\,\alpha^3$$

- 1-order splines

$$m_z(\alpha) = m_z(\alpha_k)(1-t) + m_z(\alpha_{k+1})t,$$

where $\alpha \in [\alpha_k, \alpha_{k+1}], k = 1,2,\ldots,N_k - 1,$

$$t = \frac{\alpha - \alpha_k}{h_k}, \quad h_k = \alpha_{k+1} - \alpha_k$$

- 3-order Hermitian splines, differing from classical cubic splines in that they do not require solving equations for nodal points

$$m_z(\alpha) = \phi_1(t)m_z(\alpha_k) + \phi_2(t)m_z(\alpha_{k+1}) + \phi_3(t)h_k\frac{dm_z(\alpha_k)}{d\alpha_k}(\alpha_k) + \phi_4(t)h_k\frac{dm_z(\alpha_{k+1})}{d\alpha_{k+1}}(\alpha_k),$$

where $\phi_1(t) = (1-t)^2(1+2t)$, $\phi_2(t) = t^2(3-2t)$, $\phi_3(t) = t(1-t)^2$, $\phi_4(t) = -t^2(t-1)$

and the variables are defined above.

The polynomials or splines coefficients are included into the vector of estimated parameters. To identify this vector of unknown parameters, as in the previous case, the maximum likelihood method is used [1-3]. The effectiveness of this approach strongly depends on whether the order of the polynomial or the location of the spline nodes corresponds to an unknown nonlinearity.

4. Graphic representation of nonlinearities

Identification is performed in the dialogue mode, therefore it is very important to take into account the peculiarities of the human operator perception. To improve the presentation of the results in the longitudinal channel, the following approach is proposed. From (1) - (3) it follows that estimates of the lift coefficient in a velocity coordinate system and the pitch moment coefficient for discrete points in time can be calculated as follows:

$$c_{ye}(t_i)=\frac{(n_y(t_i)\cos\alpha(t_i)+n_x(t_i)\sin\alpha(t_i))\,mg-P\sin(\alpha(t_i)+\phi_d)}{qS}, \tag{4}$$

$$m_z(t_i)=\left(\frac{J_z}{qSb_A}\right)\left(\frac{d\omega_z(t_i)}{dt}-\frac{J_x-J_y}{J_z}\omega_x(t_i)\,\omega_y(t_i)+\frac{\left(k_{\partial B}\,\omega_y(t_i)+P\,y_d\right)}{J_z}\right)$$

where ϕ_d is the installation angle of the engines, and the other symbols are explained in the description of the model (1) - (3). When using the formulae (4) the pitch angular velocity derivative we find numerically [1,6], we set the thrust according to the a priori characteristics of the engine.

Consider the widespread special case when nonlinear coefficient models have the form:

$$c_{ye}=c_{ye}(\alpha)+c_{ye}^{\delta_B}\,\delta_B,\quad m_z=m_z(\alpha)+m_z^{\delta_B}\,\delta_B+m_z^{\bar{\omega}_z}\frac{b_A}{V}\omega_z+m_z^{\bar{\dot{\alpha}}}\frac{b_A}{V}\frac{d\alpha}{dt} \tag{5}$$

The expression for the moment coefficient we write in the form

$$m_z=m_z(\alpha)+m_z^{\delta_B}\,\delta_B+(m_z^{\bar{\omega}_z}+m_z^{\bar{\dot{\alpha}}})\frac{b_A}{V}\omega_z, \tag{6}$$

where the complex of rotational derivatives is represented as one parameter [6,8].

Now we will define an approximation structure for nonlinearities (5) - (6), which in general is unknown at this stage, and perform parametric identification. Experience shows that the accuracy of estimates of linear coefficients is usually high while the nonlinearities parameters estimates are approximate. Then, using the obtained estimates of the linear coefficients and formulas (4), one can present the estimates of nonlinear dependences in an explicit form as a function of time:

$$c_{ye}(\alpha(t_i))=c_{ye}(t_i)-\tilde{c}_{ye}^{\delta_B}\,\delta_B(t_i), \tag{7}$$

$$m_z(\alpha(t_i))=m_z(t_i)-\tilde{m}_z^{\delta_B}\,\delta_B(t_i)-(\tilde{m}_z^{\bar{\omega}_z}+\tilde{m}_z^{\bar{\dot{\alpha}}})\cdot\frac{b_A}{V}\omega_z(t_i) \tag{8}$$

Next, graphs of these estimates are plotted in the function of the angle of attack, which graphically displays nonlinearity. This enables to determine the structure of nonlinearity, that is, the order of the polynomial or the location of the spline nodes.

5. Examples of non-linear dependencies identification using flight test data

The results of applying the above approach to flight tests data are presented in Figure 1. The operational range angles attack was divided into intervals (data partitioning) in which the aircraft control stick symmetrical deflections were performed so that the increments of the angle of attack were ± (2 ... 3) degrees. In figure 1, the estimates, calculated by formula (8), are shown as points, which merge into a "cloud" due to the large number of samples. It is clearly seen that at angles of attack above 11 degrees non-linear dependencies take place. Maneuvers at all 5 intervals were performed with overlapping, which allows to confirm the consistency of the results.

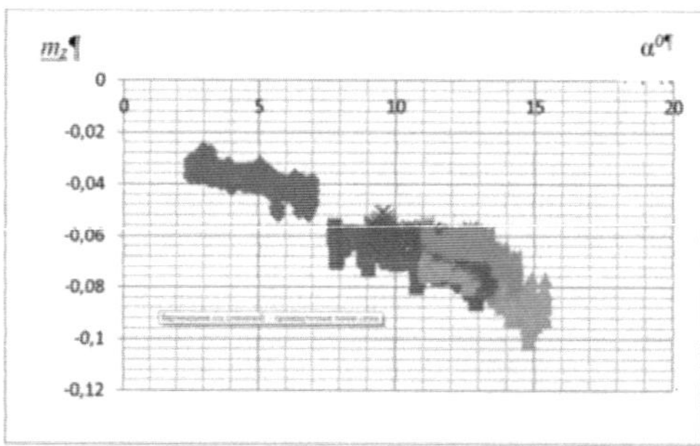

Figure 1. The results of identification of dependency when splitting into intervals.

In figure 2, for another aircraft, identification estimates of the non-linearity are shown as 5 "clouds" in comparison with the estimates obtained from wind tunnel tests (shown by large dots). In general, the degree of compliance is high, despite some differences at angles of attack of more than 14 degrees.

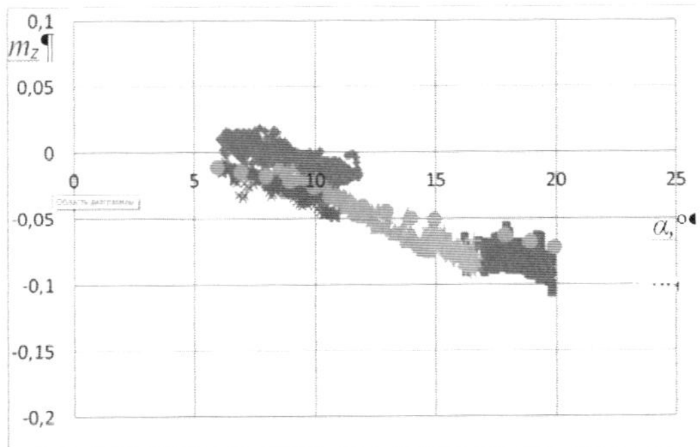

Figure 2. Comparison of the pitch moment identification estimates and the wind tunnel data.

In figure 3, the identification estimates of the lift coefficient from 5 overlapping flight intervals coincide very closely with the wind tunnel data.

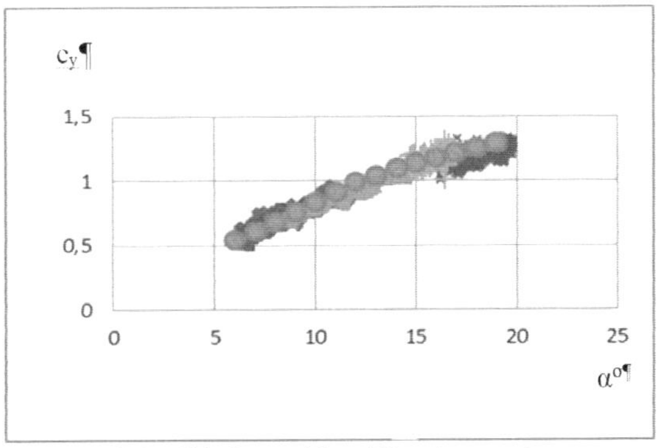

Figure 3. Comparison of the lift force identification estimates and the wind tunnel data.

Consider an example that characterizes the importance of the data presentation in the practical implementation of the aerodynamic parameters identification. In figure 4 for six flight segments, estimates of the lift coefficient by the maximum likelihood method are shown (six colored lines), in which 2-order polynomials were used to describe nonlinearity. The same figure shows the dependence taken from the wind tunnel data (indicated by the blue line). Initially, when analyzing the identification results for 6 areas, estimates of the coefficients of polynomials turn out to be significantly different. Then it is decided to display the polynomials on the same graph so that the identification estimates are displayed only for those values of the angle of attack that took place in the flight test interval. From figure 4 it is clear that this methodical technique allows us to conclude that there is a high degree of consistency between the estimates for different areas and, therefore, that the bank does not comply with the identification estimates obtained in the flight experiment. Indeed, the graphs of polynomial estimates have different slopes and displacements (that is, different coefficients of polynomials), but in general they are close to each other.

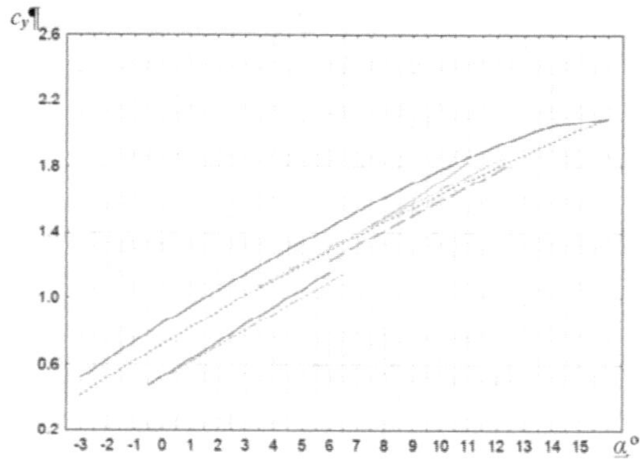

Figure 4. Comparison of polynomial lift coefficient estimates with and the wind tunnel data.

6. Conclusion

The article presents algorithmic and methodological approaches to the processing of flight data when identifying aircraft non-linear aerodynamic coefficients according to flight tests in the operational range of angles of attack. Examples of practical application of the considered methods and algorithms for identifying the aerodynamic coefficients of modern aircraft are presented. The value of the form of presenting the identification results for their correct interpretation is shown.

The presented algorithmic and methodical results make it possible to correct the mathematical models of aircraft according to flight test data.

References

[1] Klein V 1989 Estimations of Aircraft Aerodynamic Parameters from Flight Data *Progress in Aerospace Sciences* **26** (1) 1–77

[2] Klein V and Morelli E A 2006 *Aircraft system identification: Theory and practice* (Reston, VA: AIAA, Inc.)

[3] Jategaonkar R V 2006 *Flight vehicle system identification: A time domain methodology* (Reston, VA: AIAA, Inc.)

[4] Korsun O N, Nikolaev S V and Pushkov S G 2016 *Journal of computer and systems sciences international* **55**(3) 446-457

[5] Korsun O N, Stulovskii A V, Balyk O A and Zolotaeva M V 2018 *IOP Conference Series: Materials Science and Engineering* **312** 012015

[6] Korsun O N, Stulovskii A V, Ovcharenko V N and Kanyshev A V 2018 *Journal of computer and systems sciences international* **57**(3) 374-389

[7] Luchtenburg D M, Rowley C M, Lohry M W, Martinelli L and Stengel R F 2015 Unsteady high-angle-of-attack aerodynamic models of a generic jet transport *J of Aircraft* **52**(3) 890-895

[8] Bjushgens G S (Eds) 1998 *Aerodynamics, stability and controllability of supersonic aircraft* (Moscow: Nauka)

Determination of the Requirements for the Neural Network for Recognition Algorithm UHRSI

A A Ostrovskaya[1], N E Semenov[1] and A O Rubtsov[2]

[1] Peoples' Friendship University of Russia, Moscow 117198, Russia
[2] Russian Space Systems Corporation, Moscow 111250, Russia

E-mail: ostrovskaya_aa@rudn.university

Abstract. The Earth remote sensing is becoming a new and quickly developing multiscience area of a practical importance and different business applications. The IT industry now in accordance with the academic sector is developing new competences in the area which includes the development of new algorithms and calibration of existing and already working program complexes and libraries. Investigation about using neural networks for detection geo-objects on the satellite images are held by different research organizations. In this paper we present the new way of determination of the initial data, the architecture type of the neural network, the accuracy characteristics of the recognized objects for the algorithm of recognition of soil use areas of open type based on the data of ultrahigh resolution space imagery (UHRSI) which was made according to the practice on the newest satellite images. The results could be used in the various application of remote data analysis including satellite application in different branches of military and economic needs.

1. Introduction

The Earth remote sensing is becoming a new and quickly developing interdisciplinary area of a practical importance. In machine learning applications for remote sensing, aerial image interpretation is usually formulated as a pixel labeling task. Given an aerial image the goal is to produce either a complete semantic segmentation of the image into classes such as building, road, tree, grass, and water or a binary classification of the image for a single object class. So the new algorithms should be devised and studied. Below we present a way to set the preliminary conditions for the algorithm developed in the project of RUDN University and Russian Space Systems Corporation as an approach for a new competence development. Below we present our findings in determination of the requirements for the original data, the architecture of the neural network, the accuracy characteristics of the recognized objects and the development of an algorithm for recognition of quarries in accordance with ultra-high and high-resolution space imagery data.

2. Findings

1. Requirements to the initial data for the algorithm of detecting of soil use areas of open type.
   - Images from the sensor of Geoton and Wide-resolution multispectral imaging equipment of high resolution of Resource-P Satellite, panchromatic and multizonal systems of Canopus-V, Planet;

- Presence of 4 channels (red, blue, green and near infrared);
- Processing level 4A is the integrated panchromatic image (processing level 2A) and multi-spectral (processing level 2A1) images of the same territory (Pansharpening);
- Cloudiness less than 70% of the image coverage;
- The angle of deviation from the nadir of the original image should not exceed 30 degrees;
- The point sizes of each satellite image channel must be pairwise identical;
- The point sizes of the channels must have at least 1024 points in each dimension;
- The information component of each satellite channel point must be 10 bits

2. Determination of the requirements to the neural network architecture

- The neural network must have at least 10 layers;
- The neural network must have multiple inputs and one output;
- The point sizes of the result should correspond to the point sizes of the input image;
- The architecture of a neural network should provide the possibility of using methods of regularization and retraining prevention, namely, the methods of "normalizing batches" [1] and "exceptions" [2];
- The architecture should include links of the "passthrough characteristics" type [3];
- The architecture of the neural network should be adapted to the learning transfer strategy for a part of the layers

3. Determination of requirements to the accuracy characteristics of recognized objects

- The soil use areas of open type with a total linear dimension of less than 100 meters, should achieve the acquisition accuracy of 0.5 and segmentation accuracy (measure of Sørensen) of each object of no less than 0.2;
- The soil use areas of open type with a total linear dimension exceeding 100 meters, should achieve the acquisition accuracy of 0.8 and segmentation accuracy (measure of Sørensen) of each object of no less than 0.75;
- The soil use areas of open type with a total linear dimension exceeding 750 meters, should achieve the acquisition accuracy of 0.94 and segmentation accuracy (measure of Sørensen) of each object of no less than 0.8;
- The soil use areas of open type with a total linear dimension exceeding 2000 meters, should achieve the acquisition accuracy of 0.97 and segmentation accuracy (measure of Sørensen) of each object of no less than 0.95. Accuracy = TP + TN/Total (1) where TP is the number of true object definitions, TN is the number of false object definitions, and Total is the total number of unique objects. Another value 2TP/(2TP+FP+FN) (2) is the measure of Sørensen, where TP is the number of true definitions of points in the object, FP is the number of false positive definitions of points in the object, FN is the number of false negative definitions of points in the object.

3. Algorithm

Development of an algorithm for recognition of soil use areas of open type based on ultra-high resolution space imagery.

The general developed scheme of the algorithm for recognition of soil use areas of open type according to the data of ultra-high resolution space imagery is shown in Figure 1 and 2.

On entry the algorithm is awaited by 4-channel satellite image, made by canopus-B device, resurs-P, 4A processing level planet - a complex image of panchromatic (2A processing level) and multispectral (2A1 processing level) images of the same territory (Pansharpening).

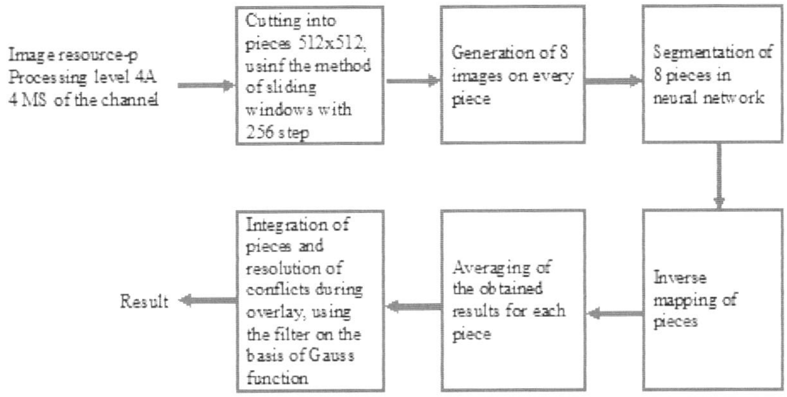

Figure 1. Diagram of the algorithm for recognition of buildings and structures

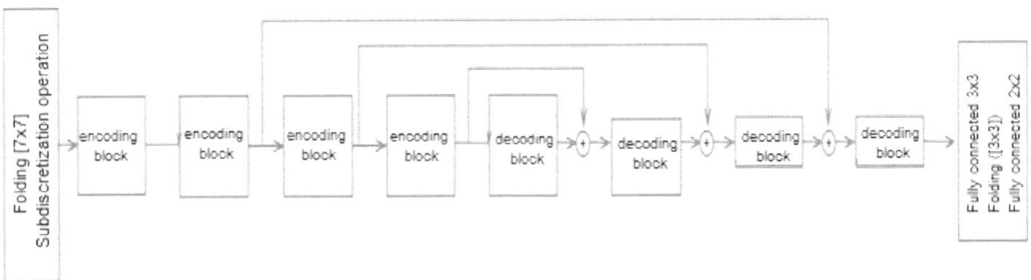

Figure 2. Architecture of the neural network

Figure 3. Splitting the original image into pieces of fixed size, using the sliding window method

Sliding window method (example on figure 3):

1. Make the size of the sliding window - $H_{SW} \times W_{SW}$ pixels;
2. Make the increments of the sliding window - S_H pixels (vertically) and S_W pixels (horizontally);
3. Complete the edges of the initial image I with the height H and width W to the size of the sliding window;
4. Form the pieces of the image (crop) of the size $H_{SW} \times W_{SW}$ from the complemented image in increments of $S_H \times S_W$

The following values were chosen for this algorithm:

– $H_{SW} = W_{SW} = 512$ pixels, to ensure the hit of rather big objects in one crop, which will allow to achieve the specified accuracy, and the practical possibility to create the software implementation on one hand, on the other hand, since the size of the window directly determines the size of the layer of the neural network, which increase leads to the increase of the requirements for the size of memory and the speed of the hardware and exponential increase of the learning time.

– $S_H = \frac{H_{SW}}{2} = 256$ pixels, $S_W = \frac{W_{SW}}{2} = 256$ pixels, to ensure the overlay of the crops so that the edge of one crop coincides with the center of the adjacent. It will enable to avoid artifacts and conflicts on the edges of the crop when restoring the segmentation of the entire image from the segmentations of certain crop areas.

Generating of 8 reflections for each crop - all possible variations of the reflection of the initial piece, using the rotation operations on π/2 and the mirror reflection (can be seen on figure 4).

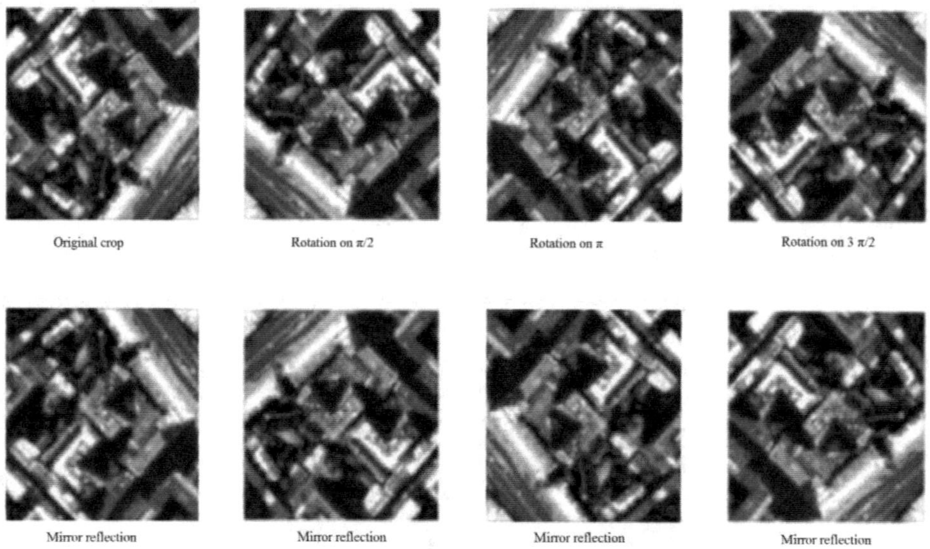

Figure 4. Turns and reflections of the crop

Let C be the crop, $\varphi(x)$ is a rotation by $\pi/2$, $\psi(x)$ is a mirror reflection, then the set 8 of reflections O can be represented as:

$$O = \{C, \varphi(C), \varphi(\varphi(C)), \varphi(\varphi(\varphi(C))), \psi(C), \psi(\varphi(C)), \psi(\varphi(\varphi(C))), \psi(\varphi(\varphi(\varphi(C))))\} \quad (3)$$

Obtain a segment map for each of the 8 reflections. At this stage, the original crop is passed through a trained deep neural web. Denote the function of obtaining the prediction P on the crop image C, as h. Then:

$$P = h(C). \quad (4)$$

The network architecture is a sequence of encoding blocks that reduce the spatial resolution of the original crop, and decoding blocks that increase the spatial resolution, combining input data with feature maps, obtained by the transmission method from the encoding blocks of the corresponding resolution, which ensures the ensemble of the results of all layers and resolutions.

The encoding block is a set of 3 operations on feature cards (figure 5).

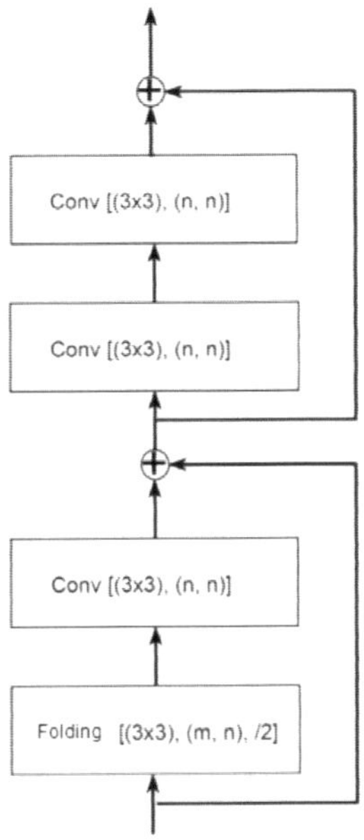

Figure 5. Block diagram of the block of encoding

These operations are:

- the convolution layer is the main block of the convolutional neural network (figure 6). The convolution layer includes its own filter for each channel, which convolution kernel processes the previous layer in fragments (summing the results of the matrix product for each fragment). It is denoted as Convolution ([k x k], m, n), where k x k is the size of the convolution kernel, m and n are the number of input and output filters for the layer, respectively;

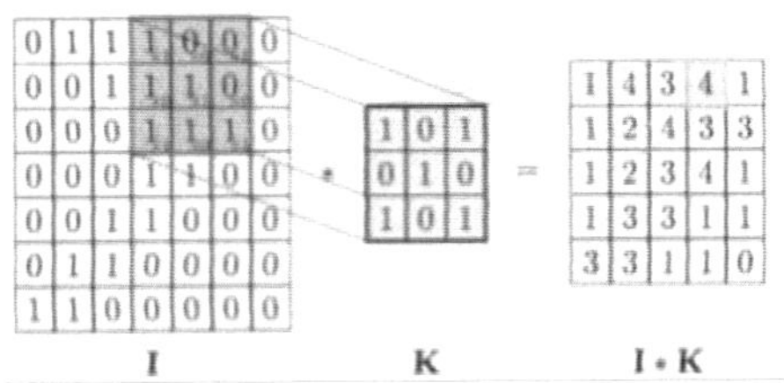

Figure 6. Example of convolutional layer operation

- ReLU activation function (figure 7). The scalar result of each convolution falls on the activation function, which is a function $\begin{matrix} += max(0,x) \\ ReLU = x \end{matrix}$ this function enables to avoid the problems of a damping and exploding gradient, and is computationally simple.

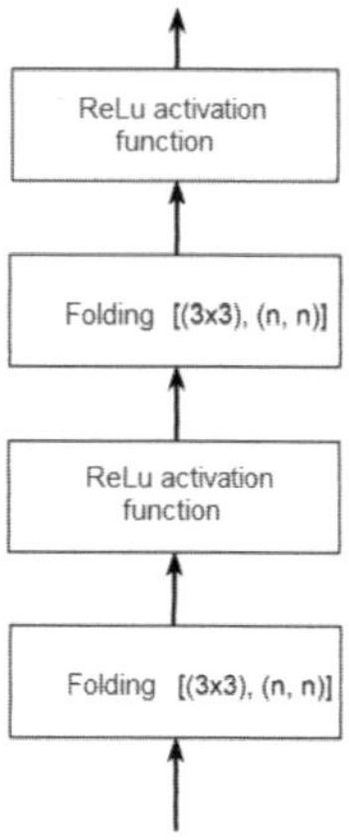

Figure 7. Diagram of the decoding unit

The decoding block is a set of 2 sequences of two operations:

- convolution layer;
- ReLU activation function.

As a result, at the output of the neural network segmentation, a set of 8 maps of probability that each point of the original crop belongs to the class "open-type subsurface resources" is obtained.

Reverse mapping operations (rotation by $-\pi/2$ and mirror mapping) are applied to the obtained set of probability maps in order to obtain the preimages of the used images with respect to the output probability maps. Since $x = \psi(\psi(x))$ and $x = \varphi^{-1}(x) = \varphi\left(\varphi(\varphi(x))\right)$, the target set O_P will have the form:

$$O_P = \{P_1, \varphi^{-1}\left(\varphi^{-1}(\varphi^{-1}(P_2))\right), \varphi^{-1}(\varphi^{-1}(P_3)), \varphi^{-1}(P_4), \psi(P_5),$$
$$\psi\left(\varphi^{-1}\left(\varphi^{-1}(\varphi^{-1}(P_6))\right)\right), \psi\left(\varphi^{-1}(\varphi^{-1}(P_7))\right), \psi(\varphi^{-1}(P_8))\} \tag{5}$$

The amendment of the boundaries of the obtained segments by averaging the predictions. Then the final prediction for each point of the crop image will be calculated, using the following formula

$$P_{result}(i,j) = \frac{\sum_{k=1}^{8} P_k(i,j)}{8} \tag{6}$$

This approach improves the segmentation results, obtained at the previous stage.

Let us combine the obtained intersecting prediction maps with the help of the weighted sum, using a two-dimensional Gaussian distribution with zero at the center of the crop and a mean-square deviation $\sigma = \frac{H_{SW}}{2*3} \approx 85$, calculated at points, corresponding to the centers of the crop pixels to get the segmentation of the original image. This will eliminate conflicts and artifacts on the crop boundaries, since the crop, which is closer to its center, will make the greatest contribution to each pixel value and the contribution of the terminal crop points is ~ 0.2.

4. Conclusion

In this paper we presented the way of determination of the initial data, the architecture of the neural network, the accuracy characteristics of the recognized objects for the algorithm of recognition of soil use areas of open type based on the data of ultra-high resolution space imagery (UHRSI) which was made according our research. The results could be used in the different application of remote data analysis. This determination is the necessary part in the development of new competences of remote data sensing [4-8]. This work lays a foundation for the new competence hub being created by RUDN and RSS, which will interact with leading Russian collectives working in remote monitoring.

Acknowledgments
This article is supported by the Program "5-100" of the Peoples' Friendship University of Russia.

References

[1] Chaurasia A and Culurciello E 2017 Linknet Exploiting encoder representations for efficient semantic segmentation *2017 IEEE Visual Communications and Image Processing (VCIP)* pp 1-4
[2] Ioffe S and Szegedy C 2015 *Batch* Normalization Accelerating Deep Network Training by Reducing Internal Covariate Shift *International Conference on Machine Learning* pp 448-456
[3] Srivastava N et al. 2014 Dropout: a simple way to prevent neural networks from overfitting *The Journal of Machine Learning Research* **15**(1) 1929-1958.
[4] Kashirin A, Semenov A, Ostrovskaya A and Kokuytseva T 2016 The Modern Approach to Competencies Management Based on IT Solutions. *JIBC-AD - Journal of Internet Banking and Commerce* **21** 1
[5] Chursin A, Kashirin A et al. 2018. The approach to detection and application of the company's technological competences to form a business-model *IOP Conference Series Materials Science and Engineering* **313** 012003 (doi:10.1088/1757-899X/312/1/012003)
[6] Kashirin A, Semenov A, Ostrovskaya A and Kokuytseva T 2016 The Modern Approach to Competencies Management Based on IT Solutions *JIBC-AD - Journal of Internet Banking and Commerce* **01** 1-12
[7] Kashirin A, Semenov A, Ostrovskaya A, Kokuytseva T and Strenaluk V 2016 The Modern Approach to Competence Management and Unique Technological Competences *Quality – access to success* **17**(154) 105–109
[8] Chursin A, Semenov A and Danilchankau A 2016 Analysis of Innovation Development in the Economy with Exhaustible Resource Sector by First Order Dynamical Systems Application *Nonlinear phenomena in complex systems* **19**(3) 254-270

IOP Conf. Series: Materials Science and Engineering **476** (2019) 012022 doi:10.1088/1757-899X/476/1/012022

Neural Net Based Multi-Agent Mobile Robots Control System: Practical Implementation on Different Platform

M L Patkin, G N Rogachev and N G Rogachev

Samara State Technical University, Samara, 443100, Russian Federation

E-mail: patkinm@gmail.com, grogachev@mail.ru and rogachev8@mail.ru

Abstract. Mobile robot control is a challenging task due to the need for lightning-fast decision making in a rapidly changing environment. We propose neural net with reinforcement learning based control system which can simultaneously manage mobile robots and communicate with other partners. The Actor Critic approach is used but we divide the Actor into the Communication Actor and Action Actor. Due to the fact that, unlike the classical approach, additional neural networks appeared, the question of optimization arose. We propose implementation of control system on three different platform: desktop, mobile and embedded platform with VPU acceleration. We demonstrate that we achieve similar performance and speed of low power embedded and mobile devices and full power desktop PC with modern GPU acceleration.

1. Introduction

The complication of technical systems and the stiffening of requirements for the accuracy of their handling involve to the need for continuous enhancement of control algorithms. In recent time, the approaches of decision making and intellectual control have enhance remarkable because of the use of artificial neural networks.

Reinforcement learning is a powerful approach for solving problems of control complicated systems, the main benefit of which is that it does not impose prior knowledge of the object model. The knowledge about the control object appears in the process of the agent's impact on the control object and receiving information about it. When solving a problem using the multi-agent approach, the complexity of the control algorithms is determined not only by the sophistication of the control objects but also by the degree of interaction of the agents. This fact suggests that practically unrestricted possibilities open up for improving the quality of facility management [1].

In this article, we propose an implementation of multi-agent cooperative interactions and a method for running this system on different platforms: PC, mobile device (iPhone) and embedded device (Raspberry Pi 3 with VPU acceleration Intel Movidius).

2. Approach

The classical method of the Actor-Critic assumes that political function of the Actor will be created on the basis of function of cost of the Critic [2, 3]. The Criticism function is developed according to an algorithm of TD [4, 5]. In cases with repeated agents (figure 1), especially in case of POMDP, cooperation between agents begins to have strong influence on quality of agents, starting with information on observation which is more important for the robot, can be found. For such situations modification of an algorithm of the Actor-Critic has been executed [6].

The main idea of the changed Actor-Critic's method for a case of the interacting systems of the multiagent in division of the Actor in the Action Actor and the Communication Actor (figure 2). The actor of action, as in a classical method responsible for development of actions of control, and the Communication Actor makes the message for partners which brings into more effective management of the system of the multiagent.

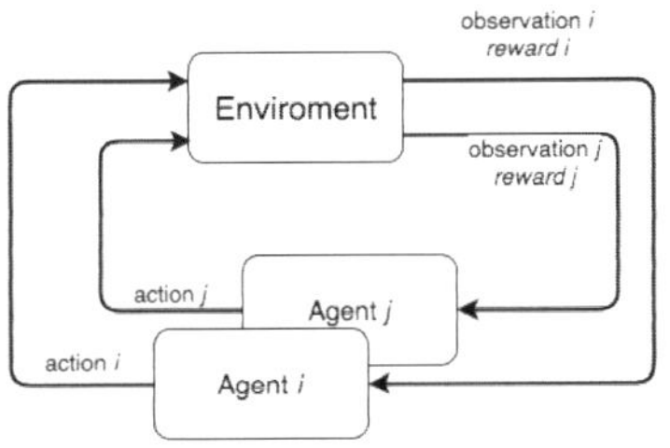

Figure 1. Multi-agent reinforcement learning.

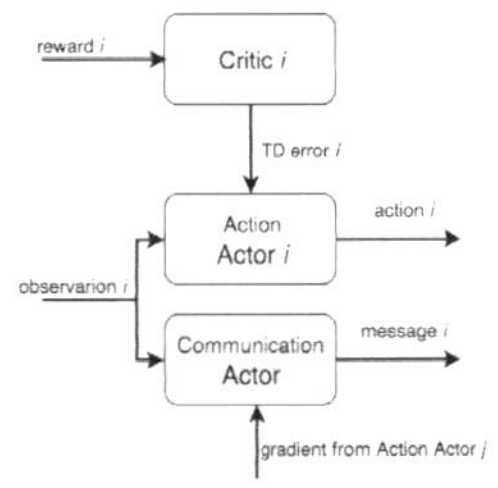

Figure 2. Scheme of modified Actor-Critic

Consider how the modified method described above works. Agents generate messages that join the observation vectors of the rest agents (Figure 3). Agents generate actions based on all information from observational vectors including information from other agents. During training and back propagation of errors from the end to the beginning, gradients propagate [7] from the Action Actor to the Communication Actor. Thus, Communication Actor is trained to generate only those messages that will reduce the number of errors in predicting the actions of Action Actor.

The figure 4 demonstrates red and green circles - agents, white lines steming from them - sonars, yellow and blue circles – targets. Table 1 demostrates architectures of neural networks of Actors and Critic where N_a – number of actions and N_m – length of message vector.

The robots are controlled by choosing one of three actions: "right", "forward", "left". The central idea of this experiment is that robot 1 knows the information about the position of target 2, but must catch target 1 and inverse. This formulation of the task allows the robots to be nested to accomplish the task, since the robots need to find a method or language of communication to accomplish it. Experiments were held for several environment settings: with a static target and with a target moving with a variable speed in an arbitrary direction.

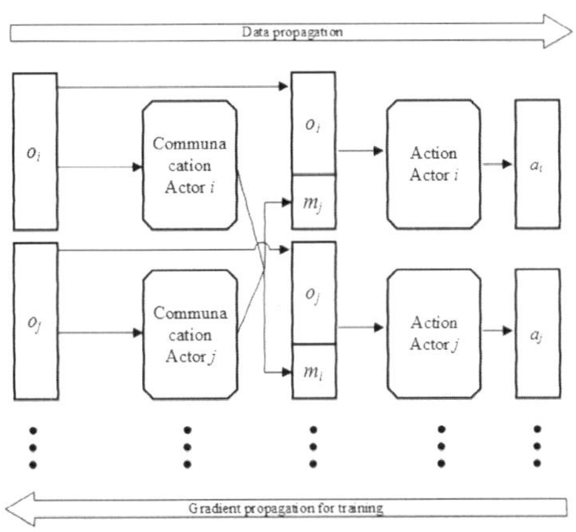

Figure 3. Modified Actor-Critic method.

IOP Conf. Series: Materials Science and Engineering **476** (2019) 012022 doi:10.1088/1757-899X/476/1/012022

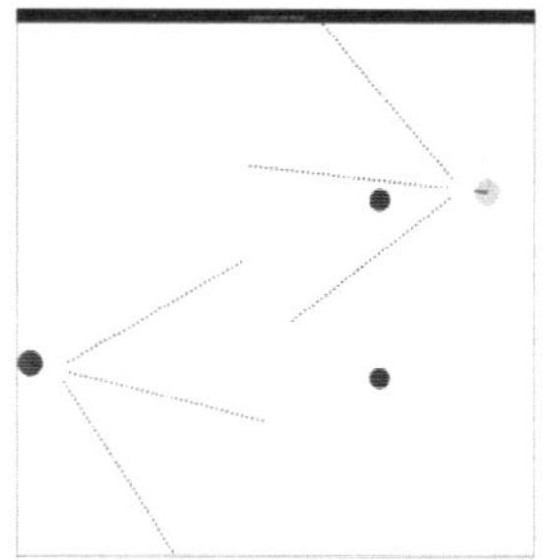

Figure 4. Visualization of environment.

Table 1. Architectures of Actors and Critic.

	Action Actor	Communication Actor	Critic
1	Fullyconnected: 1024	Fullyconnected: 1024	Fullyconnected: 1024
2	Dropout: 20%	Dropout: 20%	Dropout: 20%
3	RELU	RELU	RELU
4	Fullyconnected: 1024	Fullyconnected: 1024	Fullyconnected: 1024
5	Dropout: 20%	Dropout: 20%	Dropout: 20%
6	RELU	RELU	RELU
7	Fullyconnected: 1024	Fullyconnected: 1024	Fullyconnected: 1024
8	Dropout: 20%	Dropout: 20%	Dropout: 20%
9	RELU	RELU	RELU
10	Fullyconnected: *Na*	Fullyconnected: *Nm*	Fullyconnected: 1

3. Implementation on different platforms

3.1. Desktop
For the desktop implementation we used Keras 1.0.7 framework for the training of the models. Training of the models and environment simulation took approximately 0.5 hour on Intel i7 2600k and Nvidia Titan Black.

3.2. Mobile device
For running the trained models, we converted it to CoreML format with coremltools 2.0 which can be used on Apples' mobile devices iPhone SE and newer with operation system iOS 11.x and newer. We used iPhone 7 with A9 processor. Weights of a trained model were converted from fp32 number format (floating point format of real numbers taking 4 bytes of memory per number) to afp16(floating point format of real numbers taking 2 bytes of memory per number). This causes neglectable loss in accuracy and sufficient gain in terms of final size of a model.

3.3. Embedded device: Raspberry Pi 3 and Intel Movidius stick
For many tasks, for example when controlling robots, an embedded device is required. One of the most popular and cheap devices is Raspberry Pi 3. However, the ARM processor contained in this device is not the optimal choice for launching neural networks. Therefore, we use the Intel Movidius device to running neural networks (Figure 5).

Figure 5. Raspberry Pi 3 and Intel Movidius neural net acceleration stick

3.4. Experiments

As a result of experiments, we have shown (table 2) that neural network models work approximately the same time on both powerful PCs and mobile and embedded devices.

Table 2. Comparison of execution times of agents on different devices.

Device name	Time of execution (ms)
Desktop(i7 2600k, Titan Black)	0.7
iPhone 7	1
Raspberry Pi 3 with Intel Movidius	1.05

4. Conclusion

The suggested multiagent control method requires an increased amount of computing resources relative to standard methods, however, if you use the techniques presented in this article, you can achieve similar performance on low-power devices without loss of quality.

Acknowledgments

The reported study was funded by RFBR according to the research project № 18-08-00506A and by RFBR with Samara Region according to the research project № 17- 48-630410р-а.

References

[1] Lowe R, Wu Y, Tamar A, Harb J, Abbeel O P and Mordatch I 2017 *Advances in neural information processing systems* pp 6379-6390
[2] Konda V R and Tsitsiklis J N 2000 *Advances in neural information processing systems* pp 1008-1014
[3] Bhatnagar S, Sutton R S, Ghavamzadeh M & Lee M 2009 *Automatica* **45**(11) 2471-2482
[4] Tsitsiklis J N and Van Roy B 1997 *Advances in neural information processing systems* pp 1075-1081
[5] Sutton R S, McAllester D A, Singh S P and Mansour Y 2000 *Advances in neural information processing systems* pp 1057-1063
[6] Patkin M L and Rogachev G N 2018 Construction of multi-agent mobile robots control system in the problem of persecution with using a modified reinforcement learning method based on neural networks *IOP Conference Series: Materials Science and Engineering* **308** 012018
[7] Gupta J K, Egorov M and Kochenderfer M (2017) Cooperative multi-agent control using deep reinforcement learning *International Conference on Autonomous Agents and Multiagent Systems* pp 66-83

IOP Conf. Series: Materials Science and Engineering **476** (2019) 012023 doi:10.1088/1757-899X/476/1/012023

Genetic Algorithm and Simulation for Selecting PID-Controllers Parameters in the Walking Robot

I V Ryadchikov[1], A A Gusev[1,*], S I Sechenev[1], A S Prutskiy[1] and E Nikulchev[2]

[1]Kuban State University, Krasnodar 350040, Russia
[2]MIREA – Russian Technological University, Moscow 119454, Russia

E-mail: alexandrgsv@gmail.com, nikulchev@mail.ru

Abstract. Walking robot dynamic stabilization is a growing field of research as such robots are being constructed for various applications in human environment. The dynamic stabilization systems for such robots are often highly non-linear with the additional demand for rapid change of model mass parameters to meet the design requirements. The article centres round the problem of mechanical modelling of a walking robot and identifying the robot dynamic stabilization system parameters using the genetic algorithm approach. The stabilization system supposed in the paper is based on inverted pendulum model with paired gyroscopic units with limited kinetic momenta, that shapes out the optimization problem. The proposed control system includes a cascade of two PID regulators, the optimal parameters of which are identified using the genetic algorithm approach. The authors give the short description of the approach and suggest the fitness function suitable for this application. The computational experiment is then conducted, and the optimal parameters set is obtained. The disadvantages of the genetic algorithm solver in Simulink environment are pointed out with the approaches to their partial elimination.

1. Introduction

PID control tuning is a fairly well-studied linear programming problem [1–10]. Among the wide variety of global optimization methods [11–13] in setting up PID regulators, genetic algorithms [14–16] have received comparatively more widespread use to date, which is explained by the comparative simplicity and universality of the problem statement for genetic search , a relatively high degree of protection from “getting stuck” in the local minimum for this method and the presence of an embedded solver of genetic algorithms in lots of mathematical software packages. Nevertheless, it is necessary to note the difficulties in working with genetic algorithms. It is not always easy to translate the target requirements for a non-linear system into a formalism of a fitness function of a genetic algorithm. Additional restrictions on the parameter search area are often necessary in order to avoid obtaining solutions that lead to solver errors when simulating. Genetic search is a resource-intensive process. However, it can be speeded up using parallel computing on multi-core or multi-processor systems. These contradictions create a demand for case studying the applicability of the genetic algorithm approach to specific tasks, constructing suitable fitness functions, and describing and eliminating the difficulties encountered in the application of genetic algorithms.

This paper presents the application of a genetic algorithm to the task of tuning a cascade of two PID controllers for the walking robot body deviation compensation system. To simulate the dynamics of the robot, the model of the inverse pendulum was applied as one of the simple and at the same time adequate models [17]. Despite the above-mentioned common difficulties of using genetic algorithms,

IOP Conf. Series: Materials Science and Engineering **476** (2019) 012023 doi:10.1088/1757-899X/476/1/012023

the proposed fitness function makes it possible to find an acceptable control after 2 hours of parallel evolution on a 4-core Intel (R) Core (TM) i7-7700 microprocessor with a clock frequency of 3601 MHz and 12 GB of RAM for storing populations of parameters.

2. Formulation of the problem

The mechanical model of the walking robot consists of a beam "a" (see figure 1) of dimensions 0.02 x 0.02×2.5 m, the mass of which is negligible. The cube "body" with the side of 2 m is fixed on the beam. The center of the cube is at the height of 2.5 m (at the end of the beam). The mass of the cube is 3520 kg. Through the center of the cube passes the beam "b" perpendicular to the beam "a". Gyroscopic units GU1, GU2 are attached at the both ends of the beam "b". The maximum kinetic momentum of each of the gyroscopic units does not exceed 9000 N * m * s.

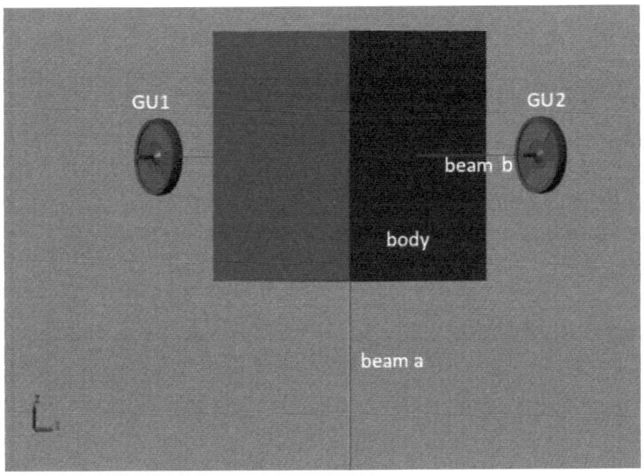

Figure 1. The mechanical model of the walking robot

When the body deviates perpendicularly to the beam "b" by 10°, it is necessary to find the optimal parameters of the PID controller cascade (see figure 2), which control the speed of antisymmetric rotation of gyroscopic units to return the robot body to the equilibrium position. For definiteness, the deviation occurs at the time t = 0. The work of friction forces is not considered.

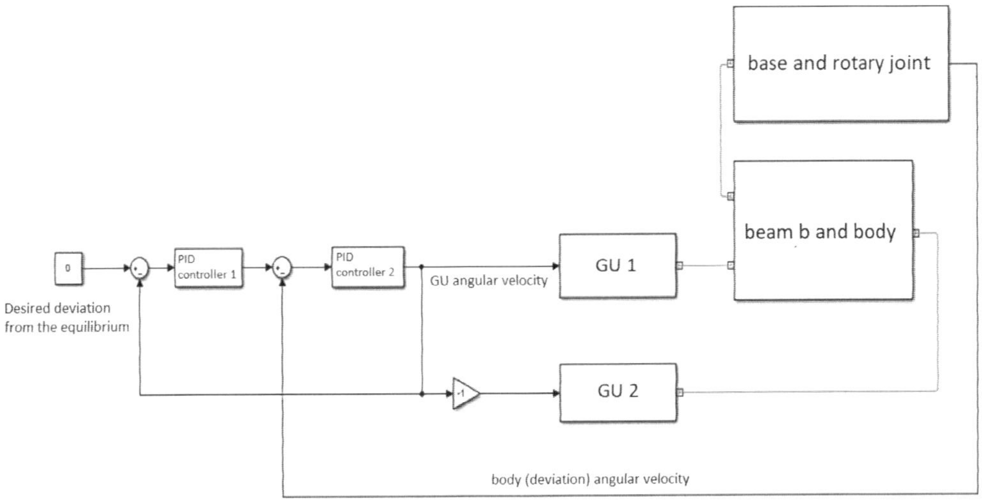

Figure 2. The proposed block diagram of the dynamic stabilization control system.

IOP Conf. Series: Materials Science and Engineering **476** (2019) 012023 doi:10.1088/1757-899X/476/1/012023

3. Method

Genetic algorithm is a global optimization algorithm based on evolutionary computations. It uses the mechanisms of random selection, combination, and variation of the desired parameters similar to the natural selection [18]. Problem formalization is carried out in the form of a genotype i.e. a vector of "genes", where each "gene" may be a bit, a number or another object. A set of genotypes of the initial population is randomly created. Each genotype receives a specific value of the fitness function. The form of the fitness function is specified by the researcher. The best genotypes in the population are selected based on their fitness values. After that the genetic operators i.e. crossing (crossover) and mutation are applied to the best genotypes, which results in a new generation of solutions for which the above-described procedure is repeated [19]. The process stops after the desired number of generations have evolved, when the allowed computational time is over, or if the best fitness does not change for a specific number of consecutive generations [20].

It should be noted that evolution may go both in the direction of increasing fitness values and in the direction of their minimization. The choice of direction depends on the software settings. The search direction may be easily inversed with the sign of the fitness function. In this paper the genetic search went towards minimizing the fitness function.

4. Solution

Using *ga* toolkit from the Global Optimization Toolbox of the Matlab R2018a the search for solution of the problem of minimizing the fitness function f was carried out. The fitness function f reflected both that robot body deviation angle is non-zero, N and that it is decreasing non-monotonously, M :

$$f = N + M ,$$

where

$$N = \sum w_1 \cdot |\varphi_i|$$

$$M = R \cdot w_2 ,$$

where R is the number of cases when $\varphi_i > \varphi_j$ provided $i > j$, φ_i is the body deviation angle for the time moment t_i, w_1, w_2 are the weight factors both equal to 100 in this study.

Optimal parameters (table 1) were found after 50 generations of parameters selection. Corresponding minimum fitness value is 2571.77. The best value of the fitness function for a randomly generated initial population of solutions was 3200.

Table 1. PID parameters found.

Controller	P	I	D	N (filter coefficient)
PID controller 1	-0.1628	-0.0439	0.0650	2.3974
PID controller 2	99.8108	143.6955	0.9534	40.5466

The dynamics of the robot body with the found parameters of the PID controller cascade is shown in figure 3.

To increase the speed of the genetic algorithm, parallel computations were used on 4 executors using the MATLAB Parallel Computing Toolbox package.

The disadvantage of the implementation of genetic algorithms in the Matlab / Simulink environment is that the solver may stuck on the parameter set corresponding to an unlimited increase in auto-oscillations in the system. The exact calculation of such oscillations requires a significant reduction in the solver time step. In some cases, the accuracy of calculations specified in the solver settings cannot be achieved with the minimum specified step length leading to the crash of the entire genetic algorithm.

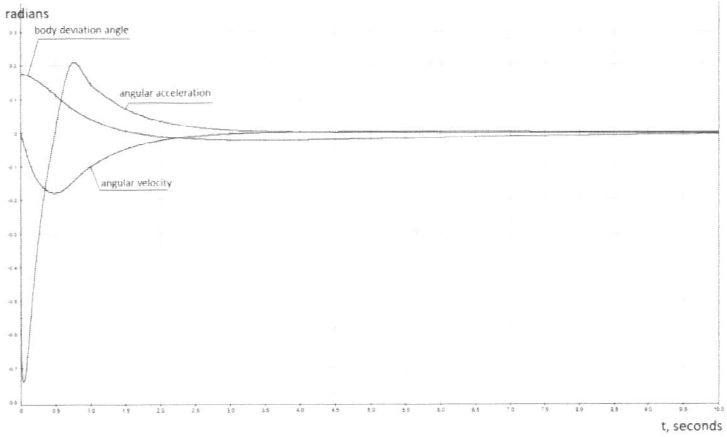

Figure 3. Robot body deviation compensation using the genetically found PID parameters.

We found two ways to deal with this drawback. The first solution is restricting the search for optimal parameters using the *InitialPopulationRange* property of the *ga* solver. However, the application of this approach can lead to finding a local minimum due to the established restrictions. In addition, it is not always possible to a priori estimate the boundaries of parameter sets that could potentially lead to difficulties for the solver.

The second way is to handle the *DiagnosticError* exception by calling the *ExecutionInfo.StopEvent* property of the metadata class of the current model simulation. In the case of such an exception, the value of the fitness function can be artificially set sufficiently large so that the genetic algorithm considers such a set of parameters undesirable and leaves the area where occur the difficulties.

5. Conclusion

The paper demonstrates the application of genetic algorithm approach for selecting the PID-controller coefficients for the dynamic stabilization system of a walking robot. A gyroscopic-based model of the robot dynamic stabilization system is proposed. The type of the fitness function to optimize the PID coefficients using a genetic algorithm is suggested. The results of the simulation of the stabilization process of the robot body using the coefficients found are presented. The results are optimal for the specified parameters of the mechanical model of the robot [21–23]. When changing the model configuration, the genetic search must be repeated using the same fitness function. Further research will be aimed at improving the dynamic stabilization system presented in the article using genetic algorithms.

Acknowledgments

The work was carried out within the framework of the state task of the Ministry of Education and Science, project No. 8.2321.2017/4.6 "Development and adaptation of control systems for compensation of dynamic deflecting effects on mobile objects in a state of dynamic equilibrium".

References

[1] Lan S and Woo P Y 1992 *Proc. IEEE Int. Conf. on Systems Engineering (Kobe, Japan)* (Hoboken, New Jersey: I.E.E.E. Press) pp 444-447
[2] Datar R, Gupta S and Langde R 2015 *IJISET* **2**(4) 948
[3] Dolezel P and Taufer I 2012 *Acta Montanistica Slovaca* **17**(3) 224
[4] Soltesz K, Mercader P and Baños A 2017 *Int. J. Robust Nonlin* **27**(11) 1857
[5] Cetin M and Iplikci S 2015 *ISA Trans.* **58** 292
[6] Yamamoto T, Fujii K and Kaneda M 1998 *IFAC Proc.* **31**(22) 59

[7] Lee C and Chen R 2015 *Sensors* **15** 11685
[8] He Y, Guo S, Shi L, Xing H, Chen Z and Su S 2017 *Proc. IEEE ICMA (Takamatsu, Japan)* (Hoboken, New Jersey: I.E.E.E. Press) p 632
[9] Niewiara L and Zawirski K 2013 *Comp. Appl. Electr. Eng.* **11** 230
[10] Güzelkaya M, Eksin I and Yeşil E 2003 *Eng. Appl. Artif. Intel.* **16** 227
[11] Coelho L S 2009 *Chaos, Solitons & Fractals* **39** 1504
[12] Ozana S and Docekal T 2016 *J. Electrical Engineering* **67** 160
[13] Carotenuto L, Pugliese P and Sergeyev Y 2006 *Control and Intelligent Systems* **34** 225
[14] Mirzal A, Shinichiro Y and Masashi 2006 *ISTECS Journal* **8** 34
[15] Sadasivarao M V and Chidambaram M 2006 *J. Indian Inst. Sci.* **86** 3434
[16] Jayachitra A and Vinodha R 2014 *Advances in Artificial Intelligence* **2014** 791230
[17] Hemami H and Golliday C L 1977 *Mathematical Biosciences* **34** 95
[18] McCall J 2005 *J. Comput. Appl. Math.* **184** 205
[19] Samakwong T and Wudhichai A 2016 *Procedia Computer Science* **86** 91
[20] Haupt R L and Haupt S E 2004 *Practical Genetic Algorithms* (John Wiley & Sons Inc.)
[21] Ryadchikov I, Sechenev S, ... and Nikulchev E 2017 *International Journal of Advanced Computer Science and Applications* **8**(9) 29
[22] Ryadchikov I, Nikulchev E, … and Feshin A 2018 *IOP Conference Series: Materials Science and Engineering* **312** 012021
[23] Ryadchikov I, Sechenev S, … and Vishnyakov R 2018 *Review of Automatic Control* **11**(4) 160

IOP Conf. Series: Materials Science and Engineering **476** (2019) 012024 doi:10.1088/1757-899X/476/1/012024

The Earliest Detection of a Gradual Change in the Properties of Random Sequences According to the Information of the Meter and the Indicator of the Accompanying Feature

G G Sebryakov, V I Pavlov, S M Muzhichek, P A Motlich, A A Skrynnikov and O V Yermolin

State Research Institute of Aviation Systems, Moscow, 125319, Russia

E-mail: msm19@yandex.ru

Abstract. There is considered a Bayesian procedure for the sequential detection of a "breakdown" of random sequences during registration together with the initial data of the accompanying attribute. The article presented that it is possible and reasonable to use additional information from non-inertia indicator for the determination of the moment of stopping and making decision of an imbalance.

1. Introduction

The task of detecting a gradual change in the properties of random sequences, or the task of detecting gradual discrepancy in process, occurs in many areas, e.g. location of moving objects, state estimation of technical devices deteriorating due to aging and wear of materials, when diagnosing the patients in medicine, etc.

It is obvious that it is necessary to detect a random moment of the discrepancy in the shortest possible time. This requirement is most satisfied with the algorithms constructed with the sequential detection methods. In the overwhelming majority of studies on the discrepancy detection [1-5], only the sequence of observed data is used without additional information. At the same time, the disorder of the observed sequence is usually associated with a change in parameters involved in the generation of data. For example, a change in the parameters of an aircraft path may be due to a change in the mode of engine operation, or a reorientation of the aircraft relative to the center of mass, or a change in the direction of the wind etc.

In many practical cases, along with tracking the informative component of the sequence it is possible to observe changes in the accompanying parameter stochastically associated with a estimated disorder in the basic data sequence. Thus, when using a radar station to determine the changing parameters of the maneuvering aircraft path, the power of the reflected radar signal will also change. Thus, in addition to the data sequence observer, it is possible to use devices which output signals are of different physical nature, but relate stochastically to the estimated disorder in the observed sequence. Such devices it is advisable to call accompanying attribute indicators. The present work on the basis of the general solution given in [6-7], solves the problem of sequential detection of gradual disassembly using the basic data sequence observer and an accompanying attribute indicator.

2. Formulation of the problem

Suppose that for a multi-step process, the output Z_n is observed,

$$Z_n = \phi_j(x_n) + W\xi_{jn}, \quad j = 0,1; \quad n = \overline{1,N}, \tag{1}$$

where x_n – source data sequence; ϕ_j – known homogeneous functions that differ from each other by the value of one of the parameters (for linear functions by the angular coefficient; for harmonic functions, either by amplitude, or frequency, or phase, etc.), $j = 0$ corresponds to the original data in the absence of disorder, a $j = 1$ – corresponds to the original data in the presence of a disorder; ξ_{jn} – white discrete centered Gaussian noises of unit intensity; W – noise intensity value; n – numbers of discrete moments of time separated by equal intervals Δt; N – the maximum number of steps of observation in one series, as well as the output signal of the indicator Y_n of the accompanying feature

$$Y_n = \Psi_{s,n}(Q), \tag{2}$$

associated with the attribute Q by the operator $\Psi_{s,n}$.

Two values of s correspond to the output signal of the indicator: $s=0$ corresponds to the absence of the accompanying attribute Q and $s = 1$ – to the presence of the attribute Q when observing the initial data x_n.

The operator $\Psi_{s,n}$ is given by the conditional probability of transition

$$\Psi_{s,n} = \Psi(s_n, n \mid s_{n-1}, Y_{n-1}, Q, n-1), \quad s = 0,1; \quad n = \overline{1,N} \tag{3}$$

from the state s_{n-1} to the state s_n [8].

When observing a Z_n sequence, there are two hypotheses: $\Theta = 0$ и $\Theta = 1$, and the hypothesis $\Theta = 1$ means a sequence mismatch that occurs at a random time k, $1 < k < \infty$ with probability $P < 1$. Here the mismatch is a gradual change in the properties of a sequence, for example, the change in the mean. We assume that we know the a priori probability of the mismatch $P_{n-1}(\Theta = 1) = P_{1,1}$ and the conditional probability densities of observations $p_0(Z_n) = p(Z_n \mid \Theta = 0)$, $p_1(Z_n \mid k) = p(Z_n \mid \Theta = 1)$ in the absence of the mismatch and its occurrence at the moment k, respectively. Observations both before and after the occurrence of the mismatch are independent, and the transition process time is shorter than the time interval Δt between successive samples [6]. It is required to make a decision on the presence or absence of a mismatch of the sequence x_n in a sequential manner, using the output signals of the observer (1) and the indicator (2).

3. Building a detection procedure

The use of additional information from the indicator is possible in two ways: the first is to control the threshold value with which a posteriori mismatch probability is compared, calculated out of the observer signal; the second is to calculate the a posteriori probabilities of hypotheses about the presence or absence of a mismatch based on the output signals of the observer and the indicator. This article discusses the first way.

Let the losses that occur when making a decision $u_n = j$ about the absence ($j = 0$) or the presence ($j = 1$) of a mismatch (discrepancy) at the n-th step have the form

$$g(\Theta, u_n, k, n) = \begin{cases} g_{0j}(n) & at\ \Theta = 0,\ n < N, \\ g_{1j}(n) + C(n-k) & at\ \Theta = 0,\ n < N, \\ \tilde{g}_{0j}(n) & at\ \Theta = 0,\ n = N, \\ \tilde{g}_{1j}(n) & at\ \Theta = 1,\ n = N, \end{cases} \tag{4}$$

where $g_{00}(n)$ – is the loss associated with the correct non-detection of a discrepancy – the cost of observations; $g_{01}(n)$ – losses caused by false alarm; $g_{11}(n)$ – losses associated with the correct detection of a discrepancy; C – losses caused by delay in detecting a disorder by one step; $g_{10}(n)$ – losses arising due to the skip of actual mismatch; $g_{0j}(N)$ and $g_{1j}(N)$ – are losses similar in content to $g_{0j}(n)$ and $g_{1j}(n)$, but related to decision making at the last step of the current series of observations and passing to the next one. In the case of gradual degradation due to the smooth change in the characteristics of technical devices for various reasons (wear, aging, changes in the temperature mode of operation, etc.), as a rule, the following ratios are true

$$\begin{gathered} g_{00}(n) \ll g_{01}(n) < g_{10}(n); \\ g_{11}(n) + C(n-k) \ll g_{01}(n) < g_{10}(n); \\ \tilde{g}_{11}(N) \ll \tilde{g}_{10}(N) < \tilde{g}_{01}(N); \\ \tilde{g}_{0j}(N) < g_{oj}(n);\ g_{1j}(n) < \tilde{g}_{1j}(N). \end{gathered} \tag{5}$$

An optimal Bayesian sequential rule for detecting a discrepancy of a Z_n sequence, $n = \overline{1,N}$, minimizing the average risk $r_n(n) = M\left[g(\Theta, u_n, k, n)\right]$, can be found by minimizing the a posteriori risk $r_n(Z_n, u_{n-1}) = r_n(Z_n)$ by choosing $j \in \overline{0,1}$. The dependence of the a posteriori risk on u_1, ..., u_{n-1} is not taken into account, since all these decisions are connected with the continuation of observations due to the absence of a discrepancy in this series.

$$r_n(Z_n) = \min\left\{ \inf_{u_n \in U_{\text{comp}}} M\left[g(\Theta, u_n, k, n) \mid Z_n, u_n\right],\ \inf_{u_n \in U_{\text{cont}}} M\left[r_n(Z_{n+1}, u_n \mid Z_n, u_n\right]\right\} \tag{6}$$

where U_{comp} и U_{cont} – areas of decisions associated with the completion and continuation of observations, respectively.

Using the formula for the total expectation, as well as the loss function (4) and its properties (5), we can determine that

$$M\left[g(\Theta, u_n, k, n) \mid Z_n, u_n\right] = \begin{cases} g_{10}(n) P_{1,n} + g_{00}(n) & at\ u_n = 0, \\ g_{01}(n)(1 - P_{1,n}) + g_{00}(n) & at\ u_n = 1, \end{cases} \tag{7}$$

where $P_{1,n} = P(\Theta = 1 \mid Z_n$ – a posteriori probability of mismatch. The solution $\delta_n(Z_n) = u_n \in U_{\text{comp}}$, at which statistics (7) takes the smallest value

$$\inf_{u_n \in U_{\text{comp}}} M\left[g(\Theta, u_n, k, n) \mid Z_n, u_n\right] = q(P_{1,n}) + g_{00}(n), \tag{8}$$

$$q(P_{1,n}) = \min\left\{g_{10}(n) P_{1,n};\ g_{01}(n)(1 - P_{1,n})\right\} \tag{9}$$

is

$$\delta_n(Z_n) = \begin{cases} j = 0 & at\ g_{10}(n) P_{1,n} < g_{01}(n)(1 - P_{1,n}), \\ j = 0 & at\ g_{10}(n) P_{1,n} > g_{01}(n)(1 - P_{1,n}). \end{cases} \tag{10}$$

The dependences similar to (7)-(10) will be valid for the second component to be minimized in (6) at the final time instant N of the current observation series

$$\rho_N(Z_n) = M\left[r_n(Z_{n+1}, u_n \mid Z_n, u_n)\right] =$$

$$= \min\left\{q(P_{1,n}) + \tilde{g}_{00}(N),\ M\left[r_N(Z_{n+1}, u_n \mid Z_n, u_n)\right]\right\},\ n = N-1, \ldots, 1,$$

$$q(P_{1,n}) = \min\left\{\tilde{g}_{10}(N)P_{1,N};\ \tilde{g}_{01}(N)(1-P_{1,N})\right\},$$

$$M\left[r_N(Z_{N+1},u_n \mid Z_n,u_n)\right] = \begin{cases} \tilde{g}_{10}(N)P_{1,n} + \tilde{g}_{00}(N) & at\ u_N = 0, \\ \tilde{g}_{01}(N)(1-P_{1,n}) + \tilde{g}_{00}(N) & at\ u_N = 1. \end{cases}$$

According to the physical meaning, $\rho_N(Z_n)$ is the smallest future a posteriori risk. The moment of termination of observations is defined as

$$\tau_{\text{comp}} = \inf\left\{n\Delta t : q(P_{1,n}) = \rho_N(Z_N)\right\}. \quad (11)$$

The optimal rule for detecting gradual mismatch can be represented as [6]

$$\delta^* = \begin{cases} u_n & at\ n\Delta t < \tau_з; \\ j=0 & at\ n\Delta t = \tau_з,\ P_{1,n} < c_з; \\ j=1 & at\ n\Delta t = \tau_з,\ P_{1,n} > c_з, \end{cases} \quad (12)$$

where the threshold value $c_з$ at the time of the termination of observations is defined as

$$c_з = g_{10}(n) / \left[g_{10}(n) + g_{01}(n)\right]. \quad (13)$$

Whatever the solutions in (12) might be, a new series of observations $n = \overline{1,N}$ begins after time (11).

The posterior probability $P_{1,n}$ of the mismatch of the Z_n sequence in the n-th step is determined on the basis of the Bayes formula

$$P_{1,n} = \left[P_{1,n-1}\Lambda_n\right] / \left[P_{1,n-1}\Lambda_n + (1-P_{1,n-1})\right],\ n = \overline{1,N},\ P_1(n=1) = P_{1,1}, \quad (14)$$

where $\Lambda_n = p_{1,n}(Z_n \mid Z_{n-1}) / p_{0,n}(Z_n \mid Z_{n-1})$.

If the indicator of the accompanying attribute is inertia-free, then we can assume that if at moment k the indicator output changes the mismatch occurs with probability $P(\Theta=1)=1$. In this case we may truncate the detection rule. The posterior probability of mismatch (14) must be compared with the new threshold value

$$c_з = \tilde{g}_{10}(N) / \left[\tilde{g}_{10}(N) + \tilde{g}_{01}(N)\right], \quad (15)$$

and the rule for detecting mismatch will be

$$\delta_k^* = \begin{cases} j=0 & at\ P_{1,k} < c_{зk}; \\ j=1 & at\ P_{1,k} > c_{зk}. \end{cases} \quad (16)$$

Comparing the values of the thresholds (13) and (15) and taking into account relations (5) we can conclude that the use of information from the indicator of the accompanying attribute leads to a reduction in the delay in determining frustration due to a lower threshold.

4. Conclusion

The procedure of detecting the gradual discrepancy in random sequences is developed in Bayesian formulation. It makes possible to reduce the delay in detection by using additional information from the indicator of the accompanying attribute. The value of the delay depends on the losses assigned by the physical sense to the wrong decisions and on the dynamic characteristics of the indicator. The indicator is considered to be non-inertia. The procedure is developed for the case of an non-controlled observation process and a low measurement cost. It is worth noting that the presence of a accompanying attribute in the source data does not necessarily mean that a discrepancy is present, and, conversely, a discrepancy may occur if the attribute is not registered. In this regard, it is advisable when analyzing long-term processes, to pre-assign the duration of a series of observations. There must be expected a definite gain in using several indicators of related traits. When using each specific indicator, due to its "correlation" with the process under study, the loss function (5), which affects the threshold (15), will be individual.

References

[1] Bakut P, Zhulina, Yu and Ivanchuk N 1980 *Detection of moving objects* (Moscow: Soviet radio) (In Russian)
[2] Maltsev A and Silaev A 1985 Detection of abrupt changes in parameters and optimal estimation of the state of discrete dynamic systems *Automation and Remote Control* **1** 48–58
[3] Kligene N and Telksnis L 1983 Methods for detecting moments of change in the properties of random processes (review) *Automation and Remote Control* **10** 5–56
[4] Sosulin Y and Fishman M 1985 *Theory of Sequential Decisions and Its Applications* (Moscow: Radio and communication) (In Russian)
[5] Shiryaev A 1996 Minimax optimality of the cumulative sum method (CUSUM) in the case of continuous time *Russian Mathematical Surveys* **51** 173–174
[6] Tartakovsky A 1987 Optimal detection of changes in the properties of random sequences. Sequential Detection *Automation and Remote Control* **7** 106–113
[7] Pavlov V 1998 Optimal detection of changes in the properties of random sequences according to the information of the meter and indicator *Automation and Remote Control* **1** 54–59
[8] Bukhalev V 1992 Recurrent algorithms for recognition and estimation of the state of a dynamic object using information from gauges and indicators *The Journal of Computer and System Sciences International* **1** 148–156

IOP Conf. Series: Materials Science and Engineering **476** (2019) 012025 doi:10.1088/1757-899X/476/1/012025

Development of High-Tech Segment in the GCC Region on the Example of the Aviation Component of the Kingdom of Bahrain

L V Shkvarya[1], I AZ Aidrous[1], E I Ruzina[1], A V Savinsky[1] and S I Rodin[1]

[1] Peoples' Friendship University of Russia, Moscow 117198, Russia

E-mail: destard@rambler.ru

Abstract. The article shows the necessity and the prospects of high-tech activities development in the GCC countries (Oman, Qatar, Kuwait, Bahrain, Saudi Arabia and the United Arab Emirates) on the example of Bahrain. It explores the role of high-tech industries in the world and in the region, in particular, the high and growing importance of aerospace activities in Bahrain and other GCC countries and substantiates the view, concerning the possibility to enhance cooperation in this area between Bahrain and Russia.

1. Introduction

Modern economic practice confirms that the level of development and socio-economic dynamics of a country and even of a whole region largely depends on the presence and degree of high-tech industries development (neither than industrialization).

The development of the high-tech sector of the economy is becoming a determining factor in economic growth, since the transition of the economy to the production of high-tech products is accompanied by a drastic decrease in the level of material consumption and energy intensity of production, productivity growth and increased competitiveness of the economy. [1] Problems of high-tech industries development in the XXI century are becoming crucial for many countries of the world.

Many authors note [2] that the modern conditions, seriously complicated by significant disproportions in the world trade force the Cooperation Council of the Arab States of the Gulf, namely Bahrain, Kuwait, Oman, Qatar, Saudi Arabia and the United Arab Emirates to strengthen the process of the national economy transformation towards non-oil specialization - both at the regional and at the global levels. Long since being a transport hub, uniting the countries of Asia, Africa and Europe, the region paid special attention to the development of the transport system. Moreover, due to different social and economic objectives of the Kingdom of Bahrain, the top priority is paid to the aviation transport.

The purpose of this article is to explore the aviation segment of the transport sector of the GCC countries in relation to the formation of new opportunities and elements of high-tech industries. This study is focused on a group of six countries and provides information on the development of air transport in the GCC countries, as well as the results of this development, which are highly differentiated with respect to the countries of the region.

To achieve this goal, the methods of structural and statistical analysis, based on the data of international organizations (UNCTAD) for 2007-2017 were used, since long-term trends allow to determine the development vector.

2. The role of high-tech industries in the global and regional economy

Modern economic practice confirms that the level of development and socio-economic dynamics of a country and even of a whole region largely depends on the presence and degree of high-tech industries development (neither than industrialization).

The development of the high-tech sector of the economy is becoming a determining factor in economic growth, since the transition of the economy to the production of high-tech products is accompanied by a drastic decrease in the level of material consumption and energy intensity of production, productivity growth and increased competitiveness of the economy [1].

A high-tech sector of the economy is most commonly understood as such a segment of the national economy in which "... high-tech technologies play a dominant, key role, and the costs on research and development (R&D) exceed the average value of this indicator in other sectors of the economy ..." [3].

At the same time, despite the long period of high-tech sectors and industries development in the modern world economy and their wide distribution both in developed and developing countries, the inclusion of a particular industry (or production) in the group of high-tech is conditional, which is confirmed by various researchers.

Thus, according to A.E. Varshavsky, "The assignment of a branch or production to the number of science-based or high-tech, accepted in foreign and domestic literature is conditional: this group includes those industries that are characterized by exceeding a certain fixed level of expenditures on R&D in relation to the amount of produced or shipped products, added value or the value of the main production factors (production assets and labor) " [5].

In this regard, approaches to the classification of high-tech industries also differ. The most widespread classification of high-tech industries is presented in Table 1.

Table 1. Approaches to the classification of high-tech industries

Item No.	According to the classification of National Science Foundation	According to the classification of UN
1.	Aviation and space rocket industry	Air and spacecraft
2.	Computers and telecommunications	Electronic computing and office equipment
3.	Electronics	Electronics, equipment for radio, television and communications
4.	Nuclear technology	Radioactive materials and other chemical products
5.	Production of weapons and military equipment	Weapons
6.	Biotechnology	Pharmaceuticals
7.	Optoelectronics	Devices (medical, optical, measuring)
8.	Development of new materials	Non-electric machines (nuclear reactors, gas turbines, etc.)
9.	Production related to computerization	Electric machines
10.	"Life Sciences"	

Source: [4]

The American National Science Foundation refers industries, including manufacturing (for example, aircraft and spacecraft; computer equipment; communication equipment and semiconductors; chemicals and pharmaceuticals; testing, measuring and control devices; motor

IOP Conf. Series: Materials Science and Engineering **476** (2019) 012025 doi:10.1088/1757-899X/476/1/012025

vehicles and their parts; railway and other transport equipment; machinery) and service sectors (for example, education, health care, business, R&D, financial and information services) to high-tech (science-based) [6].

According to the data of the annual report, Science & Engineering Indicators - 2018, issued by the foundation, industries that largely embody new knowledge and technological advances in their production, which is reflected in their R&D costs and the use of R&D in the provision of services, account for almost a third (31% in 2017) of the world economic production. The leaders in this sphere are the USA (traditionally) and China, which surpasses EU countries in this segment since 2012 (Table 2 and Fig. 1). These two countries account for 31% and 24% of the global share, respectively [6].

As can be seen from the presented data, the global crisis of 2008-2009 had a significant and negative impact on the production of high-tech products. In addition, the global recession of 2014-2015 also had a negative impact on the considered segment in the EU, Taiwan, some countries in Asia and the rest of the world as a whole.

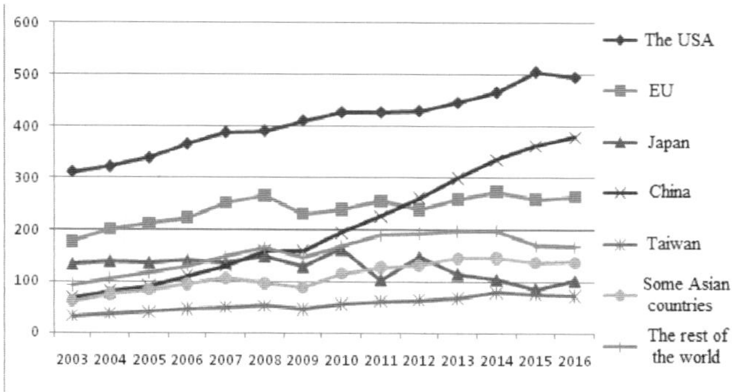

Figure 1. The release of high-tech industries' production in some countries and regions of the world in 2003-2016, billion dollars [Source: compiled, using the table. 2]

Table 2. The release of high-tech industries' production in some countries and regions of the world in 2003-2016, billion dollars.

	2003	2004	2005	2006	2007	2008	2009	2010	2011	2012	2013	2014	2015	2016
The USA	311	321	338	366	388	391	411	427	427	430	446	466	505	495
EU	177	201	211	223	252	266	230	239	255	239	259	274	259	264
Japan	133	139	136	140	136	148	129	162	103	148	115	104	86	102
China	69	80	90	110	130	158	159	194	226	262	300	336	363	380
Taiwan	32	37	40	46	49	52	46	56	61	64	68	79	76	73
Some Asian countries	61	75	83	95	106	96	88	115	127	131	145	146	137	138
The rest of the world	92	105	116	130	148	165	147	169	189	193	197	198	170	167

Source: [6]

Consequently, it can be said that high-tech segments in the economies of both countries and regions and in the global economic system strongly depend on the general socio-economic dynamics of the respective countries and regions [7].

Like the specialization model for other scientific and technical indicators, each region specializes in several different types of activities. The United States has advantages in building aircraft and spacecraft, as well as in measurement and control devices. High-tech manufacturing of aircraft and spacecraft includes a supply chain of other high-tech materials, such as navigation devices, computing and communication equipment — many of them are still provided by American suppliers.

The EU also has relative advantages in these two areas: aviation / spacecraft and measurement and control devices. China is the largest producer of a large high-tech manufacturing, information and communication technology (ICT) sub-sector, with a 34% of global share.

Despite China's rapid progress, high-tech manufacturing in the country is still largely determined by lower added value, such as a final assembly. Despite the development - and rather active - of the aerospace sphere, China still largely depends on semiconductors, supplied by foreign firms for most of its manufacturers of smartphones and other electronic devices [8].

Many experts note [9] that high-tech segments are actively developing in 'Arabian six', as well as in developing countries, such as India, South Korea and Brazil. Technology transfer processes promote their development in many ways. Experts consider technology transfer as a development factor [10].

3. Aerospace segment in the GCC on the example of the Kingdom of Bahrain

The aerospace segment in the GCC countries has been steadily and rapidly developing for the few decades. It is not only and not so much because of the accelerated growth of the transport market, both global and regional, but due to the strategic position of the region in the system of international transport flows. The steady economic growth of the countries of the region over a long period laid the financial foundation for the development of high-tech segments of the region's economies. In addition, the need to develop aviation transport and the aviation industry in general and the aerospace industry is intensified by the growing need to diversify the economies of the Cooperation Council for the Arab States of the Gulf - Bahrain, Kuwait, Oman, Qatar, Saudi Arabia and the UAE.

Worldwide, aviation creates 62.7 million jobs and brings 2.7 trillion dollars annually. In the Middle East, these figures amount to 2.4 million jobs and 157.2 billion dollars in annual regional income. According to analysts of the Boeing company, the aviation market is growing by 5% annually and by 2037 the volume of air traffic will increase by 2.5 times and the number of civil aircraft will double. In addition to orders for new aircraft, the current fleet will need to be updated by 75% in the next 20 years. Thus, the market can demand up to 40 thousand new aircraft [11].

The countries of the Gulf are keeping up with the times, investing in airport infrastructure projects, including the airport modernization program, in the formation of elements of the aerospace industry, in the development and substantial improvement of the relevant infrastructure.

Thus, Bahrain, being an island nation and a tourist center, needs air transport badly, as more than 2 million tourists, attracted by amazing antiquities, oriental culture and luxury leisure, as well as, the annual stage of 'Formula 1'. Bahrain is an important hub of international airlines.

The Ministry of Transport and Communications of the Kingdom of Bahrain is in charge of the development of the regulation of the aviation industry of the Kingdom of Bahrain. Its main goals in the considered segment are the following [12]:

1. Regulation of the civil aviation sector in order to improve flights safety and aviation security and to enhance the regulatory and supervisory role in accordance with national and international regulations.
2. Development and improvement of services in the field of navigation and meteorology in accordance with standards and international achievements.
3. Development of the infrastructure of the civil aviation sector.
4. Development of personnel potential of the civil aviation sector.
5. Work with all companies, operating in this sector to improve the quality of all services provided to all users of Bahrain International Airport.
6. Strengthening the position of Bahrain and its regional leadership in the field of civil aviation, as well as encouraging investment and employment opportunities, aimed at increasing the contribution of the civil aviation sector to the economy.

There are 4 modern airports in the country.

Bahrain International Airport, located on Muharraq Island, the largest both in this country and in Saudi Arabia, receives up to 9 million passengers a year, including Bahraini residents. It is planned to expand the airport in the next 4 years. The cost of the project, which will help to increase the number of passengers to 27 million in 30 years, 3 times more than the current capacity, is 1.8 billion Bahrain dinars. Today the Kingdom implements a number of programs to improve and modernize the infrastructure and services of the international airport and take it to the world level. Upon the project completion, the airport's capacity will increase to 14 million passengers a year.

In 2010, the Bahrain International Airport was awarded as the best airport in the Middle East, which is the proof the high quality of services. This aviation hub met such criteria as comfort, ease and speed of passenger services, etc. at the highest level.

Gulf Air, the national airline of Bahrain, is based at Bahrain International Airport. It operates flights to 39 destinations in 24 countries on three continents. The airline has one of the widest route networks in the Middle East, performing two flights a day to more than 10 cities in the region, as well as to Russia (from October 28, 2014). Moscow has become the first capital in Eastern Europe, where airline planes began flying.

Gulf Air was founded in 1950 as one of the first commercial airlines in the Middle East, so in 2015 it marked 65 years. Today, Gulf Air is a major international airline, renowned for its traditional Arabic hospitality and exceptional service quality. The Gulf Air fleet consists of 26 modern wide-body and narrow-body aircraft. The airline closely monitors market trends and constantly improves the quality of products and services for tourists and business travelers, responding to the changing needs of passengers and anticipating their expectations.

As the flag carrier of the Kingdom of Bahrain, Gulf Air is one of the key assets of the national infrastructure, which stimulates the development and growth of the state economy.

Gulf Air is an official airline and a sponsor of the Bahrain Grand Prix of the 'Formula 1' and the 'Bahrain International Airshow' in 2014.

In 2007, Bahrain Air, a private airline with a capital of 8 million Bahraini dinars (1 dollar = 0.39 dinars), will become the second national carrier in the Kingdom of Bahrain after 'Gulf Air'. 'Bahrain Air' operates flights to the Gulf countries and to Africa. At the first stage, 140 flights per week on 13 routes and the transportation of 10 thousand passengers were carried out. The company's owners are private investors from Bahrain (68% of shares) and Saudi Arabia (32%).

'Air Bahrain' is the second low-cost airline in the Gulf countries after 'Air Arabiya' (Emirate of Sharjah, United Arab Emirates).

Freight and passenger transportation by air in Bahrain is presented in the table 3.

The importance and significance of air transport for Bahrain is evidenced by the fact that since 2010, the Bahrain International Airshow has been held in the Kingdom of Bahrain, which is currently the fastest growing air show in the Middle East and attracts an increasing number of global companies. It should be noted that Bahrain can be considered as a "bridge" to the regional market in this field, as well as in many others, since the formation of a customs union in the GCC allows to cooperate with any country in the region, regardless of the country, the company has initially 'entered'.

The Bahrain International Airshow, being the fastest growing exhibition in the Middle East, reflects the steady growth of the aerospace sector of the region for years to come and will be inextricably linked with this growth as a business coordinator [13]. In 2018, it will be held from 14 to 16 November. It is interesting to note that the stand of the National Science Space Agency of the Kingdom of Bahrain (National Science Space Agency, NSSA), which was created in 2014, will be presented at the airshow.

The agency aims to create a robust infrastructure to observe outer space and the Earth, to make Bahrain achievements in space science and technology more accessible, to form a culture and technique of scientific research in the kingdom and to encourage technical innovations. In particular, it is assumed that the state will support citizens of Bahrain, working in a wide range of areas, related to space science and technology. The agency plans to launch Earth satellites for remote sensing (ERS), as well as for conducting scientific space research, including the use of the latest satellite communication technologies. Experts believe that Bahrain can benefit from the experience of such countries as Algeria, the United Arab Emirates, Saudi Arabia, Qatar, India, Singapore, Pakistan, Japan and others, having a long history in space science [14].

Table 3. Freight and passenger transportation by air in Bahrain in 2000-2015

Date	Freight		Passenger transportation	
	Volume of freight, mln ton-km	Change in accordance with the previous year, %	Number of passengers, pax.	Change in accordance with the previous year, %
2015	240,1	-7,07	5 313 756,0	2,75
2014	258,4	-2,73	5 171 496,0	11,78
2013	265,6	-21,43	4 626 648,0	-21,07
2012	338,1	-19,04	5 861 787,0	4,84
2011	417,6	-17,23	5 590 998,0	-7,26
2010	504,5	-10,6	6 028 980,0	15,6
2009	564,3	33,49	5 215 512,0	128,3
2000	208,8		1 382 461,0	

Source: compiled according to UNCTAD data.

His Excellency Kamal Bin Ahmed Mohamed, Minister of Transport and Communications and in charge of National Air Ambulance, stressed the high importance of developing the Kingdom's space science and technology sector, noting that this important work directly corresponds to the goals of Bahrain's 2030 economic vision for promoting higher education, research, innovation and government services throughout the kingdom [15].

Currently, Bahrain is negotiating with Russia for sending its astronaut into space; the UAE has already agreed. Russia can offer Bahrain cooperation in space research, the creation of advanced space technology, the provision of space services in the field of remote sensing of the Earth, space communications, navigation, space education, flights to the ISS, etc.

Bahrain can be considered as a center for the promotion of space services and products in the Gulf and Africa countries [16].

4. Conclusion

The study suggests that the development of high-tech segments of the national economy is of particular importance for the GCC countries of the XXI century. Thus, the task of creating and developing the aerospace segment as the flagship of the high-tech segment of the national economy, which the Kingdom of Bahrain is facing now, fully correlates with the national development strategy until 2030.

The GCC countries have both a need (development and diversification of the national economy in the context of the growing volatility of the global hydrocarbon market and the need to create a significant number of highly paid and prestigious jobs for the growing population of the country in the national economy) and the opportunity, predetermined by significant financial resources, generated in Bahrain and other GCC countries as a result of hydrocarbons export to the world market, to create a high-tech component and, in particular, the aerospace segment.

However, Bahrain, like other GCC countries, at this stage of activity badly need the international cooperation with the countries that can provide their assistance in achieving the ambitious plans, set by the leaders of Bahrain and the region's countries in general. In this regard, the possibility of cooperation of the Russian Federation in the aerospace sphere seems actual.

Acknowledgments

The paper was prepared with the financial support of the project №203151-0-000 «The economy of the members of Cooperation Council for the Arab States of the Gulf in the world economy digitalization»

References

[1] Gavrilova S V 2014 Conceptual Framework for the Definition of High-tech Sector of the Economy and the Operation of High-tech Companies *Economy. Statistics. Computer science* **2** 53-57

[2] Rusakovich V I 2015 Global Imbalances in the World Trade: modern approaches, problems and opportunities for regulation *Economic Systems Management.* **9(**81) 16

[3] Bendikov M A and Frolov I E 2007 *High-tech Sector of the Russian Industry: Status, trends, mechanisms of innovative development* (Moscow: Nauka)

[4] Laptev A A 2008 The Concept of "High-tech Company" in Modern Microeconomic Theory *Quality. Innovation. Education* **1**

[5] Varshavsky A E 2000 High-tech Industries and High Technologies: definition, indicators, technical policy, share in the structure of the Russian economy *Economics of Modern Russia* **2** 10

[6] *Science & Engineering Indicators* – 2018 https://www.nsf.gov/statistics/2018/nsb20181/report/sections/overview/knowledge--and-technology-intensive-economic-activity

[7] Shkvarya L, Grigorenko O, Strygin A, Rusakovich V and Shilina S 2016 *Central Asia and the Caucasus* **17** 103-112

[8] *PricewaterhouseCoopers (PwC) 2014 China's impact on the semiconductor industry: 2014 update. A decade of unprecedented growth.* http://www.pwc.com/gx/en/industries/technology/chinas-impact-on-semiconductor-industry/2014- full-report.html

[9] Rusakovich V I 2018 *Asia and Africa today* **3(**728) 48-55

[10] Solovieva Yu V 2016 *Asia and Africa today* **8**(709) 42-47

[11] *Boeing: Commercial Market Outlook* http://www.boeing.com/commercial/market/commercial-market-outlook

[12] *Ministry of Transport and Communications of the Kingdom of Bahrain* http://www.mtt.gov.bh/content/strategic-objectives

[13] *National aerospace magazine Vzlyot* http://www.take-off.ru/item/2471-aviashou-v-bakhrejne-perspektivy-razvitiya

[14] *Geospatial World* https://www.geospatialworld.net/news/bahrain-launches-national-space-science-agency/

[15] *Royal family of Bahrain: news & information* http://houseofkhalifa.com/bahrain-national-space-agency-nssa-launches-space-team-programme/

[16] Aidrous I A Z 2016 Bahrain: small country - great opportunity? *RIAC* http://russiancouncil.ru/analytics-and-comments/analytics/bakhreyn-malenkaya-strana-bolshie-vozmozhnosti/

IOP Conf. Series: Materials Science and Engineering **476** (2019) 012026 doi:10.1088/1757-899X/476/1/012026

Criterion of the Rehabilitation Process Effectiveness on the Basis of Biomehatronic System ExoLite Rehab™

S F Yatsun and A S Jatsun

South-West State University, Kursk, Russia

E-mail: teormeh@inbox.ru, ayatsun@ya.ru,

Abstract. The paper presents a biomechatronic rehabilitation system designed for a patient's active verticalization and walking with the elements of sensory integration and orthopedic correction and for locomotion training of patients with cerebrovascular pathology. An energerical criterion for the effectiveness of the rehabilitation process is proposed. The features of the construction and the methodology for the synthesis of the given motion patterns are described.

1. Introduction

Nowadays acute disorders of cerebral circulation occupy one of the leading places in terms of person mortality and disability. In Russia, the prevalence of different forms of cerebrovascular disorders is about 20% among people of working age. Stroke is an acute cerebral circulation disorder, the annual mortality from which in Russia, according to the All-Russian Center for Preventive Medicine, remains one of the highest in the world (374 per 100,000 population) [1–8].

The rehabilitation of patients, who have suffered a spinal cord injury, is also an urgent problem. Often, the negative consequences that inhibit the restoration of the body's functions are caused by the fact that a person stays at a horizontal position for a long time. Lack of mobility leads to a number of additional diseases, such as cardiovascular instability, muscle and vascular atrophy, decreased trophism, deconditioning of the body.

The success in medicine led to a shift in emphasis in approaches to this contingent from the problems of survival, to improving the quality of post-stroke patients' life, who lost their movement function. The key moment in improving the quality of a patient's life is to recover the vertical position of his body.

In this regard, the importance of using special medical technologies and technical means that can activate the functional plasticity of the brain and make the rehabilitation process more effective is growing. Locomotion training is a rehabilitation strategy designed to restore walking, the main task of which is to activate the sensory signals necessary to trigger the motion pattern of sequential neuronal excitation, that provide the contraction of the muscles involved in walking.

The appearance of exoskeletons opens new possibilities in the rehabilitation of patients and allows them to adapt to the real life conditions. In a number of countries, the work on the creation of technical devices that allow a person to move around after suffering a stroke or a spinal injury, is carried out [8-13].

As it is known, the getting up from chair and walking of the patient are the the most important and complex modes of locomotion. This motor skill, which is a chain of consistently fixed conditionally reflex motor actions performed automatically without the participation of consciousness, keeping the vertical balance, preventing the fall when moving. Automated motor act, as a result of complex coordinated activity of skeletal muscles of the trunk and extremities. When walking, support for one leg is cyclically replaced by a double-stance phase, and then supported by the other leg.

One of the approaches to rehabilitation is the application of the concept of permanent motion, alternating stops and rest. To obtain the specified patterns of motion, we offer to use lower-limb powered exoskeleton ExoLite™, which is a biomechatronic assistive-rehabilitation system that combines both a electromechanical device and the patient's body.

This man-machine device forms a synergy that provides the optimal movement of the patient's lower limbs from the rehabilitation process time point of view. Despite the large number of studies in this area, the question of using a numerical criterion for assessing the effectiveness of rehabilitation in mechanotherapy remains open. The purpose of this work is to describe and propose such criterion on the basis of the development of a rehabilitation biomechatronic system ExoLite Rehab™.

2. Mathematic modeling of the verticalization process in the exoskeleton.

The optimization of the algorithms of human movement in the exoskeleton is one of the most important tasks of biomechatronics, which has the aim to provide a highly effective rehabilitation process by implementing the necessary movement patterns that will minimize the recovering time. It is very important to have objective criteria for the effectiveness of the rehabilitation process. To formulate this criteria, let us consider the work performed by the human muscles and the electric drives of the exoskeleton in the process of the patient standing up.

To develop estimation methods of the effectiveness of the human rehabilitation process, it is proposed to measure the parameters characterizing the patient's standing up process. From the point of view of kinematics, this kind of motion is the simplest and, at the same time, the most power consuming, so the activity of the patient's muscles in this process can be estimated quite easily. To estimate power inputs, we will consider the scheme of the person in an exoskeleton which is shown in figure 1.

The following symbols are used in the scheme: φ_2, φ_3, φ_4 – the angles defining the orientation of links 2, 3 and 4 with respect to the horizontal; $\boldsymbol{C_i}$ – the center of mass of the ith link; C – the center of mass of the exoskeleton; O_{i+1} – the joint connecting the ith link with the next one; M_i – the torque of the electric drive and the muscles of the patient acting in the ith joint (i=1, 2, 3). The links have masses m_i and lengths l_i. Let us assume that the links are absolutely rigid bodies and the foot of the exoskeleton (link 1) remain motionless during verticalization.

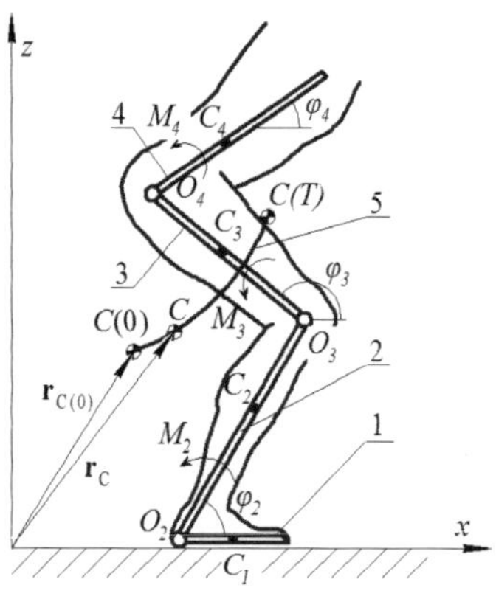

Figure 1. Scheme of a patient in exoskeleton standing up: 1-4 – links of the exoskeleton; 5 – the trajectory of the center of mass; $\overline{\mathbf{r}}_C$ – the current position of the center of mass of the exoskeleton; $\overline{\mathbf{r}}_C(0)$ – the position of the center of mass at the initial moment of time.

Let us introduce the vector of generalized coordinates $\bar{q}=[\varphi_2 \quad \varphi_3 \quad \varphi_4]^{\mathrm{T}}$, which determines the mechanism position.

The energy of muscles and electric drives is expended to compensate of the work performed by forces of gravity and resistance. For example, when performing verticalization of the patient with the help of the exoskeleton from the sitting position during the time T, it is necessary to carry out work on changing the vertical coordinate of the center of mass of the man-exoskeleton system.

This work can be determined by the formula:

$$A_{mg} = \int_0^T mgV_{Cz}dt, \tag{1}$$

where A_{mg} – mechanical work on changing the position of the vertical coordinate of the center of mass of the man-exoskeleton system; m is the mass of the man-exoskeleton system; g – acceleration of gravity; V_{Cz} is the vertical velocity of the center of mass of the man-exoskeleton system; t – current time; T is the time of the verticalization process.

And also it is necessary to perform work for compensation resistance forces which can be defined by the following formulas:

$$A_C = \sum_{i=1}^{n}\int_0^T M_{Ci}\cdot\dot{\varphi}_{i,i-1}dt, \tag{2}$$

where A_C is the mechanical work of resistance forces, M_{Ci} – resistance moment in the ith drive; $\dot{\varphi}_{i,i-1}$ – angular velocity in the ith joint; n – number of drives in the exoskeleton.

The resistance moment in each drive M_{Ci} depends on the magnitude and type of friction arising in friction pairs, including bearings, gears and other elements of the drive.

With total disability of the patient musculoskeletal system, the muscles work is equal zero. Thus, the work on lifting up the patient is performed only by exoskeleton drives and consists of two components:

$$A_e = A_{mg} + A_C \tag{3}$$

where

$$A_e = \sum_{i=1}^{n}\int_0^T M_{ie}\dot{\varphi}_{i,i-1}dt = \eta_e\sum_{i=1}^{n}\int_0^T u_i I_i dt, \tag{4}$$

U_i is the voltage supplied to the ith drive; I_i is the winding current of the ith drive; n is the total number of drives; η_e – efficiency of the drive.

If the muscles of the patient's legs get into the process of rehabilitation and perform part of the work by the exoskeleton, then the work on lifting up the patient's body will consist of two components:

$$A = A_e + A_m = A_{mg} + A_C, \tag{5}$$

where A_e — the work of the exoskeleton and A_m — work of human muscles.

In this case, to determine the work **A** lifting the patient's body next formula is applicable:

$$A = \sum_{i=1}^{n}\int_0^T (M_{ie} + M_{im})\dot{\varphi}_{i,i-1}dt, \tag{6}$$

where M_{im} – the moments generated by the muscles at the corresponding joints.

We'll find a work produced by the muscles. From (5) we get

$$A_m = A_{mg} + A_C - A_e \tag{7}$$

The parameter determined by the (7) allows to assess the condition of the patient's muscles during rehabilitation.

$$\frac{A_m}{A_e} = \frac{A_{mg}}{A_e} + \frac{A_C}{A_e} - 1. \tag{8}$$

Convert (8) together with (5) to the following view:

$$\eta = \frac{A_m}{A_e} = \frac{A}{A_e} - 1. \tag{9}$$

where η is the relation of the work performed by the patient, to the work performed by the exoskeleton actuators. Work values A and A_e are easily measured by well-known methods.

The offered dimensionless criterion gives us possibility to quantify the condition of the patient muscles before and after rehabilitation. Thus it is possible to estimate numerically the effectiveness of the rehabilitation process.

3. Description of the rehabilitation system

Biomechatronic rehabilitation system ExoLite Rehab™ is designed for active patient verticalization, walking and exercises with feed-back sensor integration and orthopedic correction. It is a universal training tool for patients with walking disabilities caused by diseases and injuries of the central nervous system [13-15].

In fact, it is an individual device for patients with various forms of motive disorders. The rehabilitation method carried out by low-intensity simultaneous sensory stimulation (tactile, vestibular, kinesthetic receptors, mechanoreceptors of the feet) combined in time with correction of movements in the form of difficulty of stereotyped movements, obstructing the adoption of reflex-conditioned postures, facilitating physiological movements in a vertical position.

The main advantage of this device over analogues is that the rehabilitation doctor or trainee can independently simulate and implement a variety of movement patterns for the patient, taking into account the highlights of disorders and individual human characteristics.

Robotic system ExoLite Rehab™ allows patients with motive disorders of the lower limbs to perform basic movements such as standing, walking, flexion and extension in some joints (hip, knee, ankle) with different amplitude and rate of exercise [16, 17].

ExoLite Rehab system based on lower limb exoskeleton equipped with a power frame, patient fixation system, drives that ensure the movement of the patient's limbs according to the law given by the doctor, sensors that allow measuring the rotation angles in the patient's joints in real time and assess the patient's condition.

During patient verticalization provides the stimulation of the internal sensory inputs (tactile, vestibular, proprioceptive systems) and integration of these systems with the formation of the "body scheme" and the posture-tone control.

A distinctive feature of the complex issue of the formation of normal movement patterns is the implementation of the beginning of any motion of a symmetrical body position with a moderately stiff fixation of the pelvic girdle and waist. That meets the requirements of the Bobath concept therapy [18, 19]. The basis of the Bobath concept are: understanding of normal and pathological stereotypes of movement; maximum level of independence in everyday life; much attention is paid to the restoration of impaired sensitivity.

The weakening of pathological synergies and synkinesia increases the ergonomics of movements. The fixation system provides support and fixation of the spine along the entire length, which is necessary in the case of scoliotic deformation and lumbar hyperlordosis, as well as physiological position, unloading and moderate adjustable fixation of the hip, knee and ankle joints.

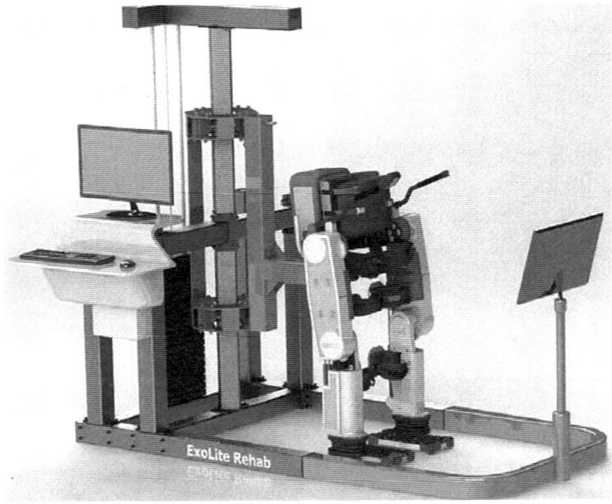

Figure 2. General view of the rehabilitation complex ExoLite Rehab™

The complex is equipped with an intelligent control system with an intuitive two-channel interface that allows the doctor and the patient to set the necessary commands. To ensure the comfortable conditions of the patient, the system of fixing cuffs has adjustments that allow for personal adjustment of the complex for each patient and to carry out rehabilitation most effectively.

During moving the patient in a corrected position, while facilitating the physiological movements and the difficulty of pathological, the following mechanisms are implemented: conversion of primitive tonic reflexes at anti-gravity, dynamic proprioceptive correction, the destruction of pathological synergies, the formation of new reflex connections.

Decrease spasticity, strengthen muscles, increase range of motion, increase depth of breathing, activation of the reactions of straightening and balance, improving ergonomic in routine activities lead to a general increase in mobility increases the ability to self-care and communication.

Rehabilitation is carried out in the following modes:

1. A slow walk is 1.5-2.5 km/h. With this speed it is recommended to move at the initial stage of rehabilitation after injuries and strokes.
2. Medium-speed walking 3-4 km/h. On the body of people with heart defects and a weakened cardiovascular system, such the pace of walking has a beneficial effect.
3. It is possible to achieve fast pace-4-5 km/h.
4. Periodically, in the verticalization mode, the muscle activity of the patient is measured using the criterion (9).

The rehabilitation complex can be used in the body mass index close to normal (26-27), with the preserved function of the upper limbs, in the period of the disease from 6 months to 4 years, with any degree of violation of conductivity in ASIA.

4. Definition of the motion control law

The strategy of exoskeleton motion control is based on two fundamentally different approaches. On the one hand, it is a program control of the links movement along specified trajectories, which are determined by the rehabilitation doctor, and on the other hand - it is copying control, in which the patient can change the position of the exoskeleton links himself.

At the first stage of rehabilitation, when the functional activity of the muscles is small or absent, it is advisable to apply the program control. Let us describe how the defining effects are formed to implement the verticalization process with "ExoLite Rehab".

To do this, consider the problem of determining the rotation angles of the exoskeleton links depending on the given motion of the center of mass.

It is necessary to determine the angles that determine the position of the exoskeleton links in space, that is, to find a vector of generalized coordinates

$$\bar{q} = (\varphi_2, \varphi_3, \varphi_4)^{\mathrm{T}} \tag{10}$$

corresponding to the specified trajectory of the center of mass of the exoskeleton at each time.

In this case, the inverse kinematics problem is solved. Let a vector $\bar{F}(\bar{q})$ that determines the coordinates of the center of mass C depending on $\bar{q}$. In this case we will get

$$\bar{F}(\bar{q}) = r_C^*(\bar{q})$$

$r_C^*(\bar{q})$ – radius-vector of the center of mass of the man-exoskeleton system.

Vector $\bar{q}$ has the restrictions associated with the physiological characteristics of the musculoskeletal system and design of the exoskeleton power frame:

$$\bar{q}^- < \bar{q} < \bar{q}^+,$$

where $\bar{q}^-$ – the minimum value of the vector; $\bar{q}^+$ – the maximum value of the vector.

We also introduce a vector function that defines the position of point C on a given trajectory as a function of time:

$$\bar{\Phi}(t) = [x_C^*(t), y_C^*(t), z_C^*(t)]^{\mathrm{T}}. \tag{11}$$

It is obvious that

$$\bar{F}(\bar{q}) = \bar{\Phi}(t) \tag{12}$$

The resulting equation (12) allows us to determine the values of the vector $\bar{q}$ for a given function $\bar{\Phi}(t)$.

In discrete form the solution looks like:

$$\bar{q}^{k+1} = \bar{q}^k + \alpha_\kappa J^{-1}\Delta\bar{\Phi}(t) \tag{13}$$

where $\Delta\bar{\Phi}(t)$ – function increment in the time step Δt; $J^{-1} = (\frac{\partial \bar{F}}{\partial \bar{q}})^{-1}$ – inversed Jacobian matrix.

The obtained relation allows us to find the vector of generalized coordinates $\bar{q}$ at the $(k+1)^{\mathrm{st}}$ time step according to the known value at the kth step. This vector sets the values of absolute angles for each link of the exoskeleton and is further denoted as $\bar{q} = \bar{q}^*$.

5. Experimental results

The experimental study of the process of verticalization was conducted on biomechatronics complex "ExoLite Rehab".

As a patient, a robotic dummy was used, which allows to simulate the interaction of the patient and the rehabilitation complex, including simulating muscle activity, and the value of the criterion can vary in the range: $\eta = 0 - 0.3$.

Fig.3 shows the nominal exposure derived using the algorithm (13) in the form of dependences of the absolute angles φ_2, φ_3, φ_4 for a given trajectory of the center of mass , constructed from the condition that the point of the ZMP (zero moment point) is inside the supporting polygon (a stop of the exoskeleton), which are implemented by electric drives. At fig.4 time dependences of the torques of "ExoLite Rehab" electric drives are presented with rehabilitation criterion $\eta = 0$. Fig. 5 shows time dependence of the total torque of the actuators and artificial muscles dummy $\eta = 0.1$ and $\eta = 0.25$ for the case when the muscles of the patient partially restored their activity after rehabilitation .

IOP Conf. Series: Materials Science and Engineering **476** (2019) 012026 doi:10.1088/1757-899X/476/1/012026

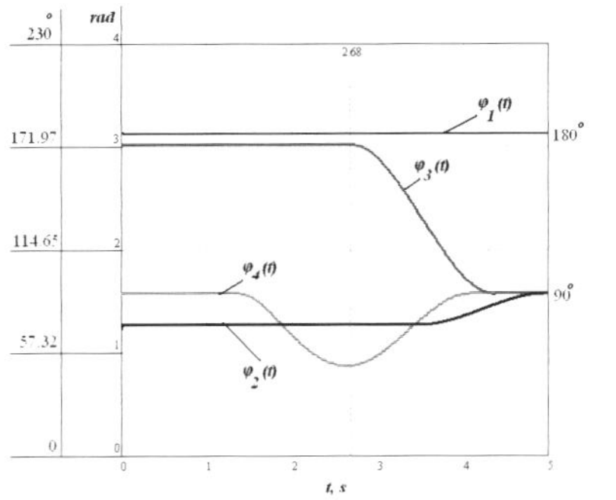

Figure 3. Time dependences of the absolute angles φ_2, φ_3, φ_4

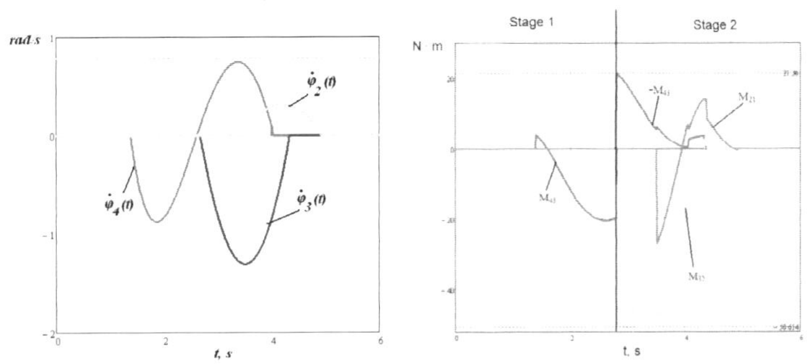

Figure 4. Chart change of the angular velocity and the control torques of the actuators. The dummy's muscles do not work

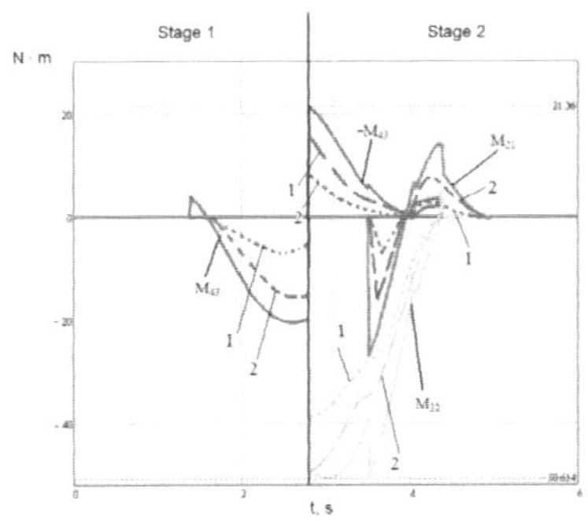

Figure 5. Time dependence of the total torque of the actuators and artificial muscles

6. Conclusions

Improving the effectiveness of rehabilitation measures is possible through the use of high-tech equipment that allows the rehabilitation doctor to provide individual approaches based on the selection of unique patterns of movement of the lower leg, the patient's thigh with optimal amplitude and frequency, which opens up opportunities for the implementation and personified approach in the treatment of the patient.

A new generation of equipment has been developed and manufactured, which has the ability to deeply regulate all parameters of motion and, what is especially important, allows to realize the measurement of the energy spent by the patient. An objective criterion for the effectiveness of the rehabilitation process is proposed. To formulate this criterion, the work performed by human muscles and exoskeleton actuators in the process of patient verticalization is considered.

Acknowledgments

Research supported by Russian Foundation of Basic Research project #18-08-00773A.

References

[1] Ama D, Antonio J, Gil-Agudo Ángel, José Pons L. and Moreno C J 2014 *Journal of neuroengineering and rehabilitation* **11**(1) 27
[2] Diedam H, Dimitrov P B, Wieber K, Mombaur M and Diehl D 2008 *IEEE/RSJ International Conference on InIntelligent Robots and Systems, IROS 2008* pp. 1121-1126
[3] Dumitru N, Copilusi C, Geonea I, Tarnita D and Dumitrache I 2015 *New Trends in Mechanism and Machine Science* (Springer Intern. Publ) pp 709-717
[4] Khairul A and Ali Al-Jumaily A 2012 *Procedia Engineering* **41** 988-994
[5] Talaty M, Esquenazi A and Briceño J E 2013 *2013 IEEE 13th International Conference on Rehabilitation Robotics (ICORR)* pp. 1–5
[6] Steger R, Kim S H and Kazerooni H 2006 *Proceedings 2006 IEEE International Conference on Robotics and Automation, ICRA 2006* pp 3469-3476
[7] Zoss A B, Kazerooni H and Chu A 2006 *IEEE/ASME Transactions On Mechatronics* **11**(2) 128-138
[8] Wallis S and Bernie M 2017 A treatment proposal with lower extremity robotic devices: Performance of postural analysis and treatment with REXERCISES *2017 International Symposium on Wearable Robotics and Rehabilitation (WeRob)*
[9] Dhand S, Singla A and Virk G S (2016) A brief review on human-powered lower-limb exoskeletons *Conference on Mechanical Engineering and Technology (COMET-2016), IIT (BHU)* (*Varanasi, India)*
[10] Diedam H, Dimitrov P B, Wieber K, Mombaur M and Diehl D 2008 *IEEE/RSJ International Conference on InIntelligent Robots and Systems, IROS 2008* pp. 1121-1126
[11] Dumitru N, Copilusi C, Geonea I, Tarnita D and Dumitrache I 2015 *New Trends in Mechanism and Machine Science* (Springer) pp 709-717
[12] Jatsun S, Savin S and Yatsun A 2016 Improvement of energy consumption for a lower limb exoskeleton through verticalization time optimization *2016 24th Mediterranean Conference on Control and Automation (MED)* pp 322-326
[13] Jatsun S, Savin S and Yatsun A 2016 Comparative analysis of iterative LQR and adaptive PD controllers for a lower limb exoskeleton *2016 IEEE International Conference on Cyber Technology in Automation, Control, and Intelligent Systems (CYBER)* pp 239-244
[14] Jatsun S, Savin S, Yatsun A and Malchikov A 2016 Study of controlled motion of exoskeleton moving from sitting to standing position *Advances in Robot Design and Intelligent Control* (Springer) pp 165-172
[15] Jatsun S, Savin S, Yatsun A and Turlapov R 2015 Adaptive control system for exoskeleton performing sit-to-stand motion *2015 10th International Symposium on Mechatronics and its Applications (ISMA)* pp 1-6
[16] Vereykin A A, Savchenko A G et al 2015 *Engineering Bulletin* (4) 1-11
[17] Vorobiev A A, Andryushchenko F A et al 2015 *Volgograd Scientific Medical Journal* (3) 14-20
[18] Savin S I, Yatsun S F, Yatsun A S and Vorochaev L Yu 2015 *Proceedings of the 11th All-Russian Congress on Fundamental Problems of Theoretical and Applied Mechanics (Kazan)* p 244
[19] Yatsun S F, Pavlovsky V E, Lushnikov B V et al 2015 *Exoskeletons: analysis of structures, classification, principles of creation, modeling basics* (University Book, Kursk: SWSU)

Institute of Physics
Dirac House, Temple Back
Bristol BS1 6BE UK

ISSN: 1757-8981
ISBN 978-1-5108-8217-1

CURRAN ASSOCIATES INC.
proceedings.com